A PHOTOGRAPHIC ATLAS OF CAT ANATOMY:

With Sheep Heart, Brain, Eye, Kidney Dissection

THIRD EDITION

MICHAEL J. TIMMONS
Moraine Valley Community College
Biology/Health Sciences Division
Palos Hills, Illinois

Spyglass Publications
Orland Park, Illinois

Notice: The author and the publisher of this volume have taken care that the information contained herein is accurate and compatible with the standards generally accepted at the time of publication. The author and the publisher disclaim any liability, loss, or damage incurred as a consequence, directly or indirectly, of the use and application of any of the contents of this volume.

Third Edition
ISBN 0-9627559-2-3

Printed in the United States of America

Spyglass Publications
8242 Spyglass Circle
Orland Park, Il 60462

ISBN 0-02-4208441-8
Published by:
Macmillan Publishing Company
866 Third Avenue
New York, N.Y. 10022

ABCDEFGHIJ

Library of Congress Catalog Card Number: 97-062351

PREFACE

Because cat anatomy is similar to that of the human, the cat is often used as a dissection specimen in undergraduate classes where a cadaver is not available. An essential element in courses in mammalian anatomy, comparative vertebrate anatomy, and anatomy and physiology is the dissection and examination of laboratory specimens. This atlas provides anatomy students with a guide to the essentials of cat anatomy. This book is not meant to be a complete treatise on the subject of cat anatomy, but is rather, a visual primer designed to help students in the systematic observation of cat anatomy and mammalian organs. The photographs in this atlas are intended to supplement the dissection descriptions found in laboratory manuals and are comprehensive enough to provide the basis for individualized instruction.

Because most laboratory manuals and atlases usually contain only line drawings that often lack depth, students find that their specimens do not closely resemble the drawings. The photographs selected to appear in A Photographic Atlas of Cat Anatomy were carefully chosen to show actual specimens at different stages of dissection. This series of 69 photographs shows the step-by-step dissection of the cat in realistic photographs that depict what students actually will see during dissection. All pertinent anatomical structures are clearly labeled; photographs progress sequentially from topographical views to detailed dissections in close-up views. The terminology used is the conventional nomenclature of the Nomina Anatomica Veterinaria as adopted by the World Association of Veterinary Anatomists in 1973.

New to this edition are photographs depicting the dissection of the sheep heart, brain, spinal cord, eye, and kidney. Photo figures have been reorganized in the third edition so that the dissection sequence follows the same course system sequence as in most undergraduate vertebrate anatomy and physiology courses. Additional cat structures have been labeled and the index has been expanded to include a greater number of anatomical structures.

I wish to acknowledge and express my sincere appreciation to the following persons for their help in the preparation of this atlas: my students, for their encouragement; Thomas J. McCague, Professor, Biology/Health Sciences Division at Moraine Valley Community College for reviewing the manuscript; and a special note of thanks to Mary M. Oreluk, RN for her valuable suggestions, diligent typing and proofreading.

I extend a note of appreciation to my family for their interest and support in the publication of the third edition of this book, and special thanks to Judy, for her constant help and encouragement.

M. J. Timmons
Orland Park, Illinois
January, 1998

LIST OF FIGURES

CIRCULATORY SYSTEM - HEART **FIGURE**

CIRCULATORY SYSTEM - BLOOD VESSELS

RESPIRATORY SYSTEM - LUNGS AND THORACIC VISCERA

ABDOMINAL VISCERA - DIGESTIVE SYSTEM

UROGENITAL SYSTEM - KIDNEY

UROGENITAL SYSTEM - FEMALE

UROGENITAL SYSTEM - MALE

A NOTE TO THE STUDENT

When an anatomical term is introduced for the first time, the term is accompanied by phonetic pronunciation. For quick and easy reference, anatomical terms are alphabetically listed in the index by system, with the Figure number appearing next to the term.

To assist you in integrating the photographs with your instructional sequence, a list of all photographs that appear in the book may be found in the List Of Figures. The dissection photographs are arranged and listed by body system. This approach allows flexibility in rearranging or omitting certain areas of dissection. Each photograph is captioned to indicate the area being dissected and the view of dissection, and provides visually the information necessary to guide you in your dissection of a particular anatomical area.

Anatomical variations do occur in individual specimens and, as a result, these structural modifications may not always exactly resemble a particular photograph. The student should keep this in mind, particularly when studying the blood vessels and nerves, because their lengths and locations vary slightly from specimen to specimen.

Viewing each of the photographs of a specific body system prior to actual dissection will help to introduce and orient you to the specific region to be studied. Use the photos as a visual reference during dissection while you systematically examine and study each pertinent anatomical structure. When properly used, this Atlas will reduce the amount of your time spent in the laboratory on the mechanics of dissection, thereby leaving more time for observation and examination.

Quadruped animals like the cat are referenced in the horizontal position; whereas, humans are referenced in the vertical (standing) position. The following terms are used in the Atlas to designate mammalian anatomical position:

Anterior (Cranial):	Head end of the animal.
Craniad:	Toward the head or anterior end of the animal.
Caudal (Posterior):	Tail or posterior portion of the animal.
Caudad:	Toward the tail or posterior portion of the animal.
Dorsal:	Back of the animal.
Dorsum:	Back of the animal or upper portion of a limb.
Lateral:	Away from the midline of the animal.
Medial:	Toward the midline of the animal.
Ventral:	The underside of the animal or of a limb.
Distal:	Part or appendage farthest from the body or point of reference.
Right and Left:	Always refer to the specimen's right or left.

Sections of the Body

Sagittal:	A longitudinal section passing from anterior to posterior through the median long axis of the body.
Parasagittal:	A sectional plane that lies on either side of the sagittal section.
Transverse:	A section that cuts the body from dorsal to ventral into anterior and posterior portions.
Frontal (Coronal):	A section that cuts the body horizontally from anterior to posterior into dorsal and ventral portions.

FIGURE 1 CAT SKELETON - LEFT LATERAL VIEW

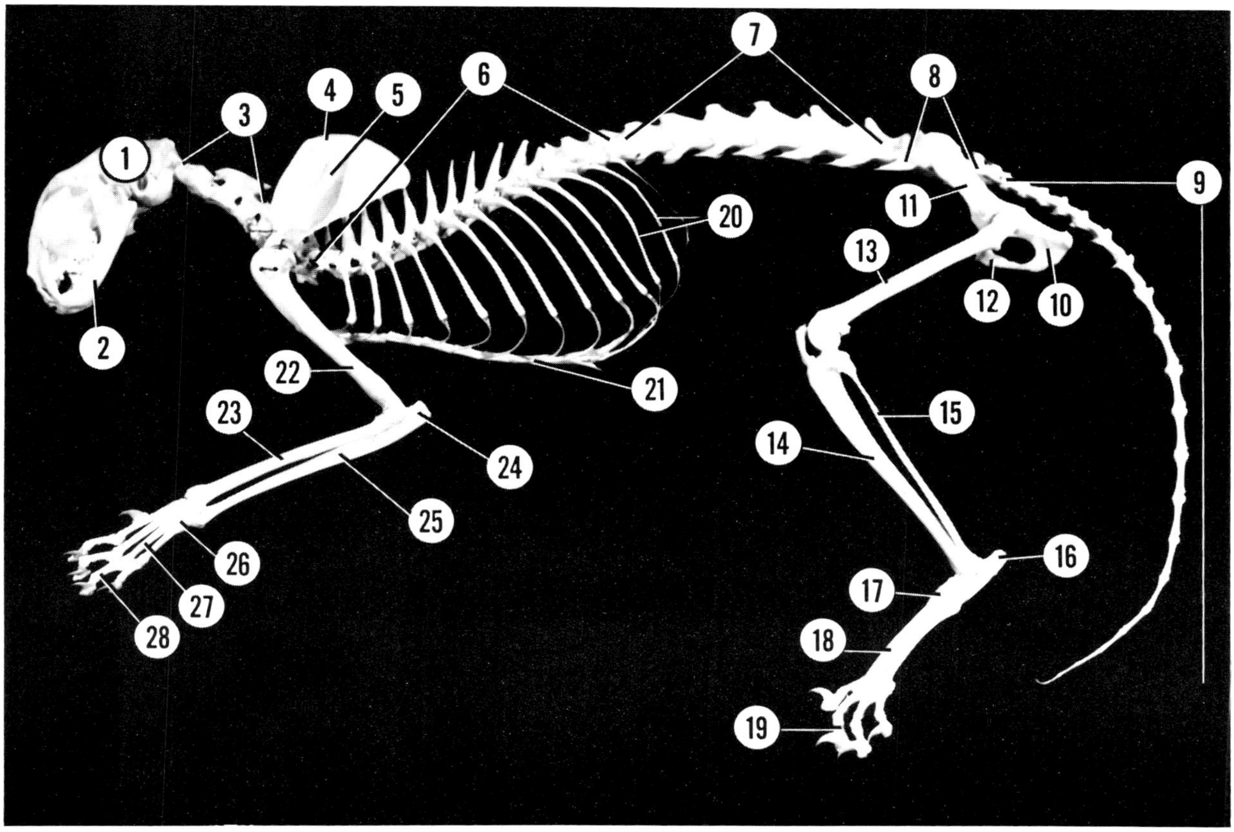

1. **Skull**
2. **Mandible** (MAN-di-bal)
3. **Cervical vertebrae** (SER-vi-kul VUR-te-brā)
4. **Scapula** (SKAP-yoo-la)
5. **Scapular** (SKAP-yoo-lar) **spine**
6. **Thoracic** (thō-RAS-ik) **vertebrae**
7. **Lumbar** (LUM-bar) **vertebrae**
8. **Sacrum** (SĀ-krum)
9. **Caudal** (KAW-dal) **vertebrae**
10. **Ischium** (IS-kē-um)
11. **Ilium** (IL-ē-um)
12. **Pubis** (PYOO-bis)
13. **Femur** (FĒ-mur)
14. **Tibia** (TIB-ē-a)
15. **Fibula** (FIB-yoo-la)
16. **Calcaneus** (kal-KĀ-nē-us)
17. **Tarsals** (TAR-suls)
18. **Metatarsals** (met´ a-TAR-suls)
19. **Phalanges** (fa-LAN-jēz)
20. **Ribs**
21. **Sternum** (STUR-num)
22. **Humerus** (HŪ-mer-us)
23. **Radius** (RĀ-dē-us)
24. **Olecranon** (ō-LECK-ra-non) **process**
25. **Ulna** (UL-na)
26. **Carpals** (KAR-puls)
27. **Metacarpals** (met´ a-KAR-puls)
28. **Phalanges**

FIGURE **2**

A VENTRAL PANORAMIC VIEW OF THE CHEST, ABDOMEN AND APPENDICULAR MUSCLES AS SEEN AFTER REMOVAL OF THE SKIN

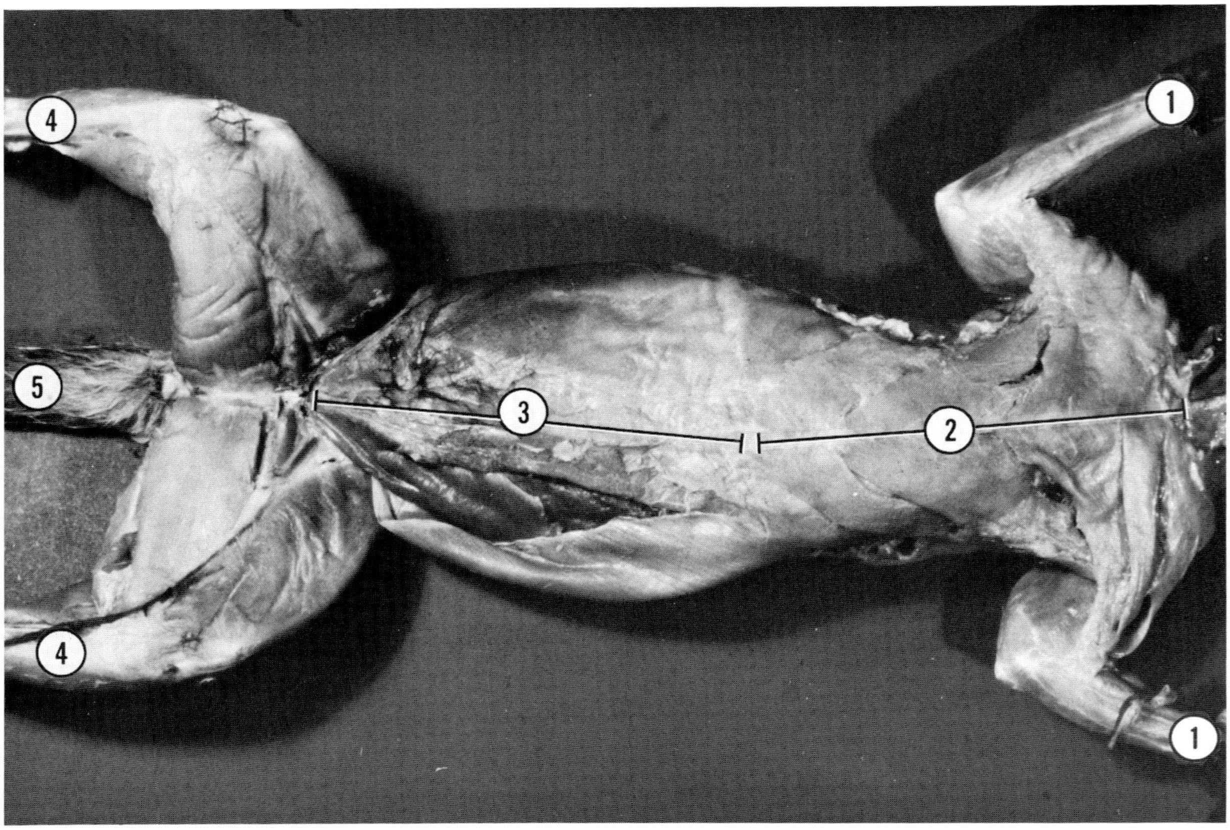

1. **Forelimbs**
2. **Chest**
3. **Abdomen**
4. **Hindlimbs**
5. **Tail**

FIGURE 3 SUPERFICIAL CHEST MUSCLES - VENTRAL VIEW (LEFT SIDE)

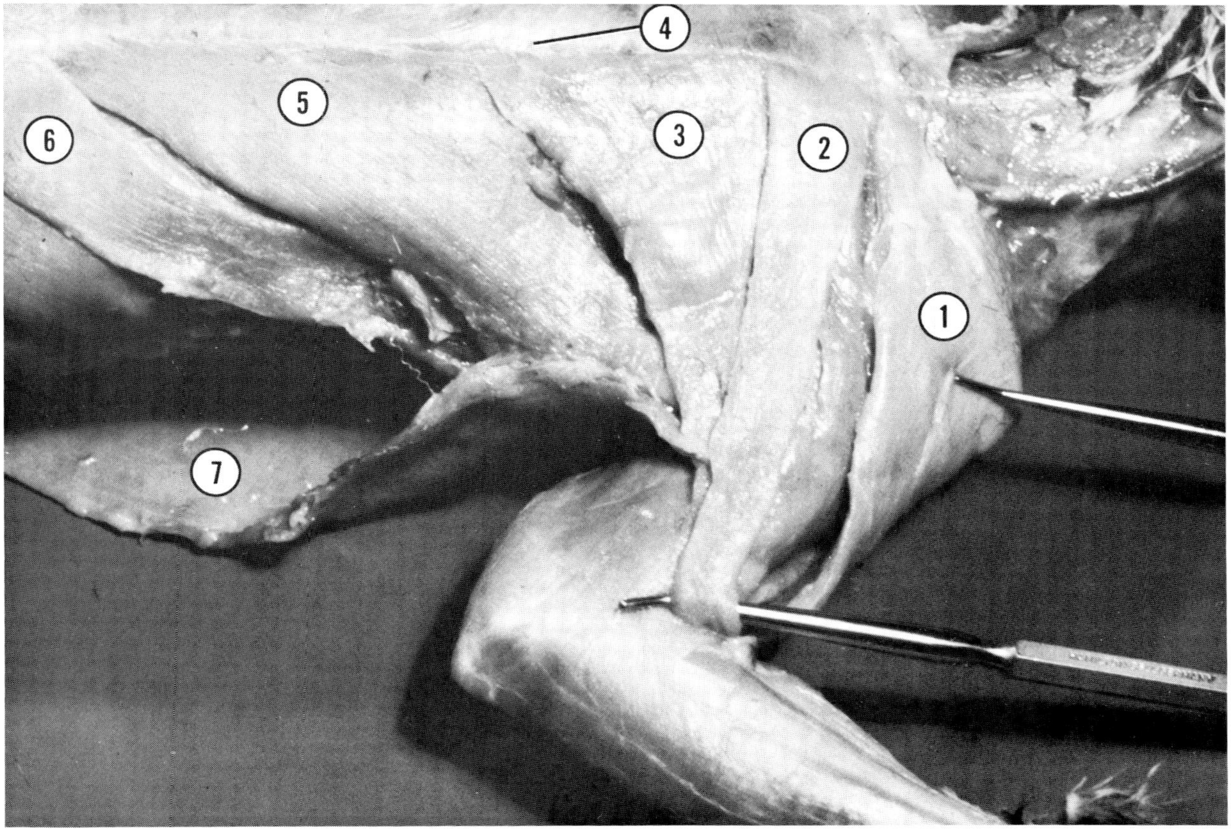

1. **Clavodeltoid** (klav-ō-DEL-toyd) **or Clavobrachialis** (klav´ ō-brā-kē-Ā-lis)
   - **A:** Flexes forelimb
   - **O:** Clavicle
   - **I:** Ulna, proximal end
2. **Pectoantibrachialis** (pek´ tō-an-tē-brā-kē-Ā-lis)
   - **A:** Adducts forelimb
   - **O:** Sternum, manubrium end
   - **I:** Forelimb, proximal end
3. **Pectoralis** (pek´ tō-RĀ-lis) **major**
   - **A:** Adducts humerus
   - **O:** Sternum
   - **I:** Humerus, shaft portion
4. **Linea alba** (LIN-ē-a AL-ba)
5. **Pectoralis minor**
   - **A:** Adducts humerus
   - **O:** Sternum
   - **I:** Humerus, bicipital groove
6. **Xiphihumeralis** (zif-ē-hyoo´ mer-AL-is)
   - **A:** Adducts humerus
   - **O:** Sternum, xiphoid process
   - **I:** Humerus, bicipital groove
7. **Latissimus dorsi** (la-TIS-i-mus DOR-sē)
   - **A:** Dorsal and posterior retraction of humerus
   - **O:** Spinous process of T_4 -L_6 vertebrae
   - **I:** Humerus, shaft

FIGURE **4** **CHEST AND ABDOMINAL MUSCLES - LATERAL VIEW (LEFT SIDE)**

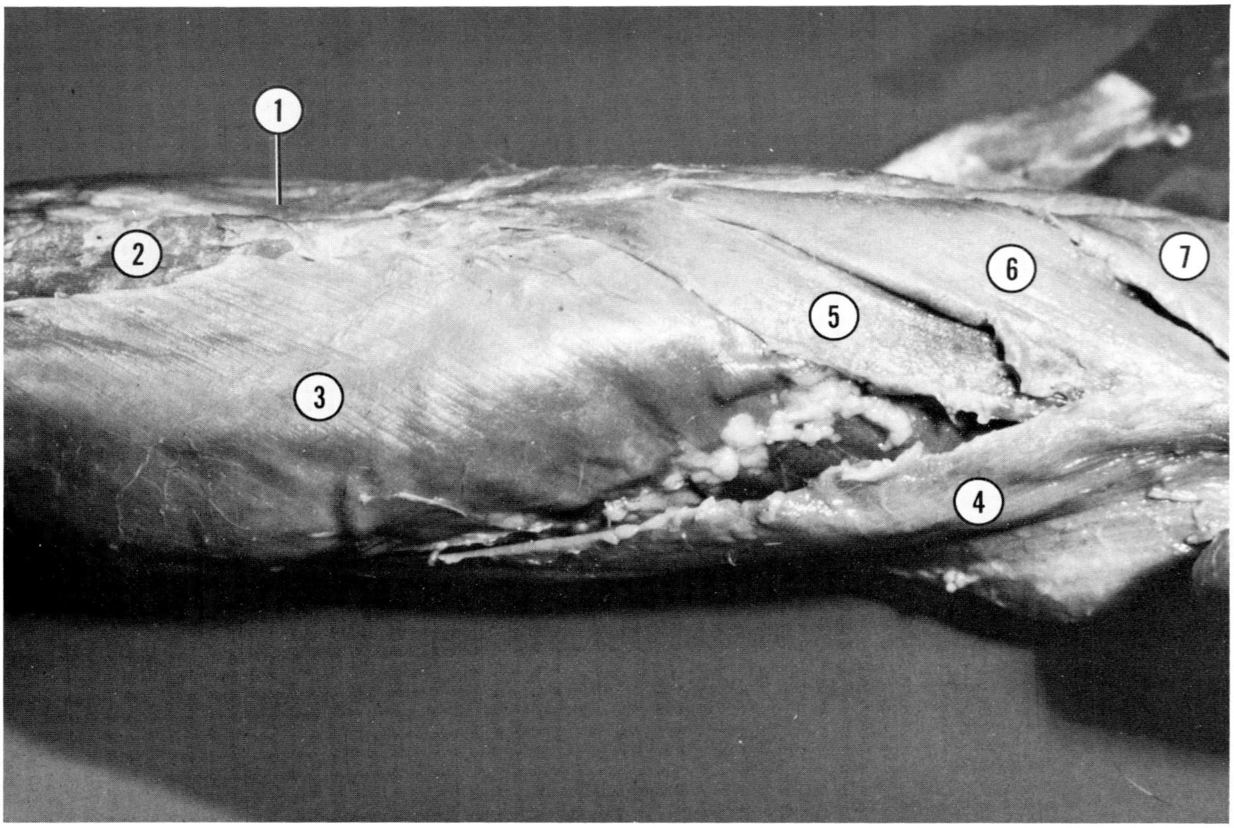

1. **Rectus abdominis** (REK-tus ab-DOM-i-nus)
   - **A:** Flexes trunk
   - **O:** Sternum
   - **I:** Tubercle of pubis
2. **Transversus abdominis**
   - **A:** Constricts abdomen
   - **O:** Ilium, diaphragm, transverse lumbar processes, false & free ribs
   - **I:** Linea alba
3. **External abdominal oblique** (ō-BLĒK)
   - **A:** Constricts abdomen; lateral bending
   - **O:** Ribs; last ten
   - **I:** Linea alba and pubis
4. **Latissimus dorsi** (la-TIS-i-mus dor-sē)
5. **Xiphihumeralis** (zif-ē-hyoo´ mer-AL-is)
6. **Pectoralis minor**
7. **Pectoralis major**

FIGURE 5 DEEP MUSCLES OF THE CHEST - VENTROLATERAL VIEW (LEFT SIDE)

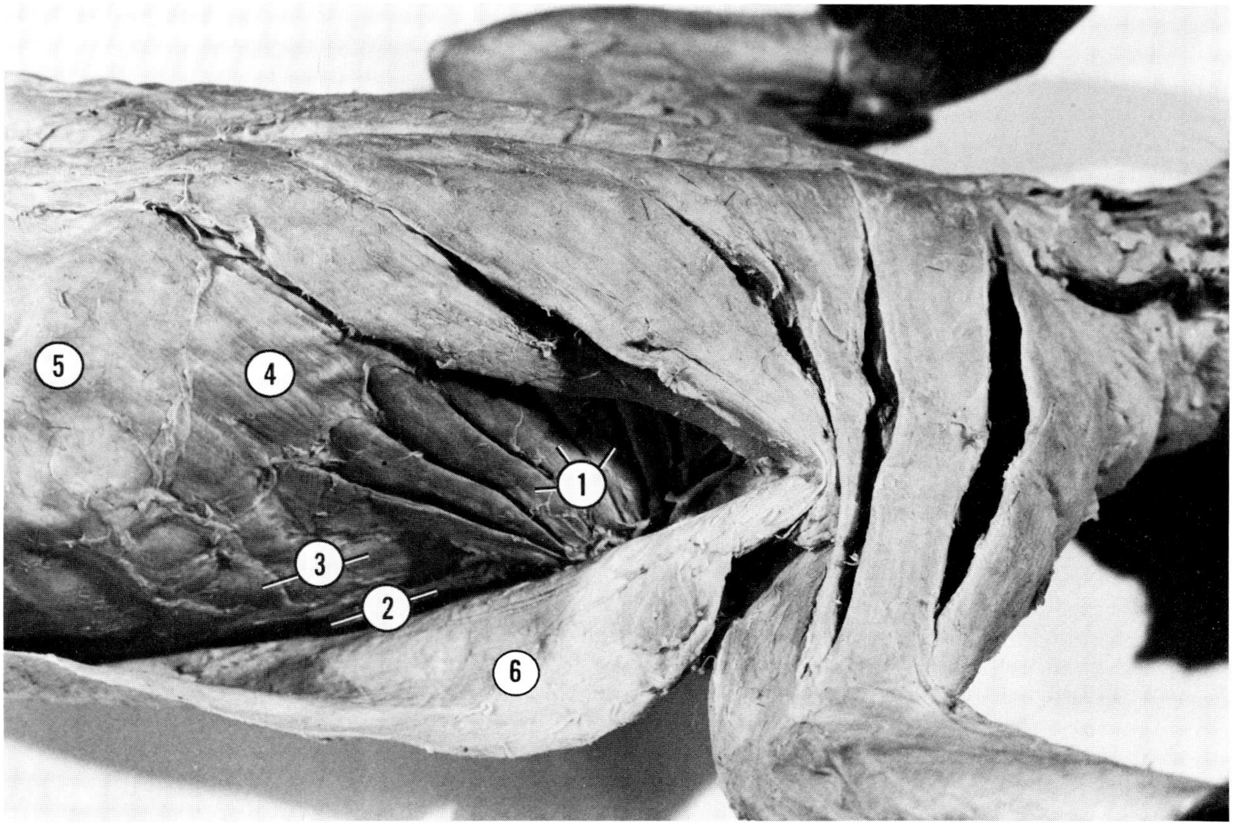

1. **Serratus ventralis** (se-RĀ-tus ven-TRAL-is)
   - **A:** Depresses scapula
   - **O:** Ribs, 1 to 10
   - **I:** Scapula, vertebral border
2. **Scalenus medius** (skā-LĒ-nus MĒ-dē-us)
   - **A:** Flexes neck, moves ribs craniad
   - **O:** Ribs, 6 to 9
   - **I:** Cervical vertebrae, transverse processes
3. **Serratus dorsalis caudalis** (dor-SĀ-lis KAW-dal-is)
   - **A:** Assists in expiration
   - **O:** Lumbar vertebrae, spinous processes
   - **I:** Ribs, 9 to 13
4. **External intercostals** (in´ ter-KOS-tulz)
   - **A:** Extends ribs during inspiration
   - **O:** Ribs, posterior border
   - **I:** Ribs, anterior border
5. **External abdominal oblique**
6. **Latissimus dorsi**

FIGURE 6 ABDOMINAL MUSCLES - VENTROLATERAL VIEW

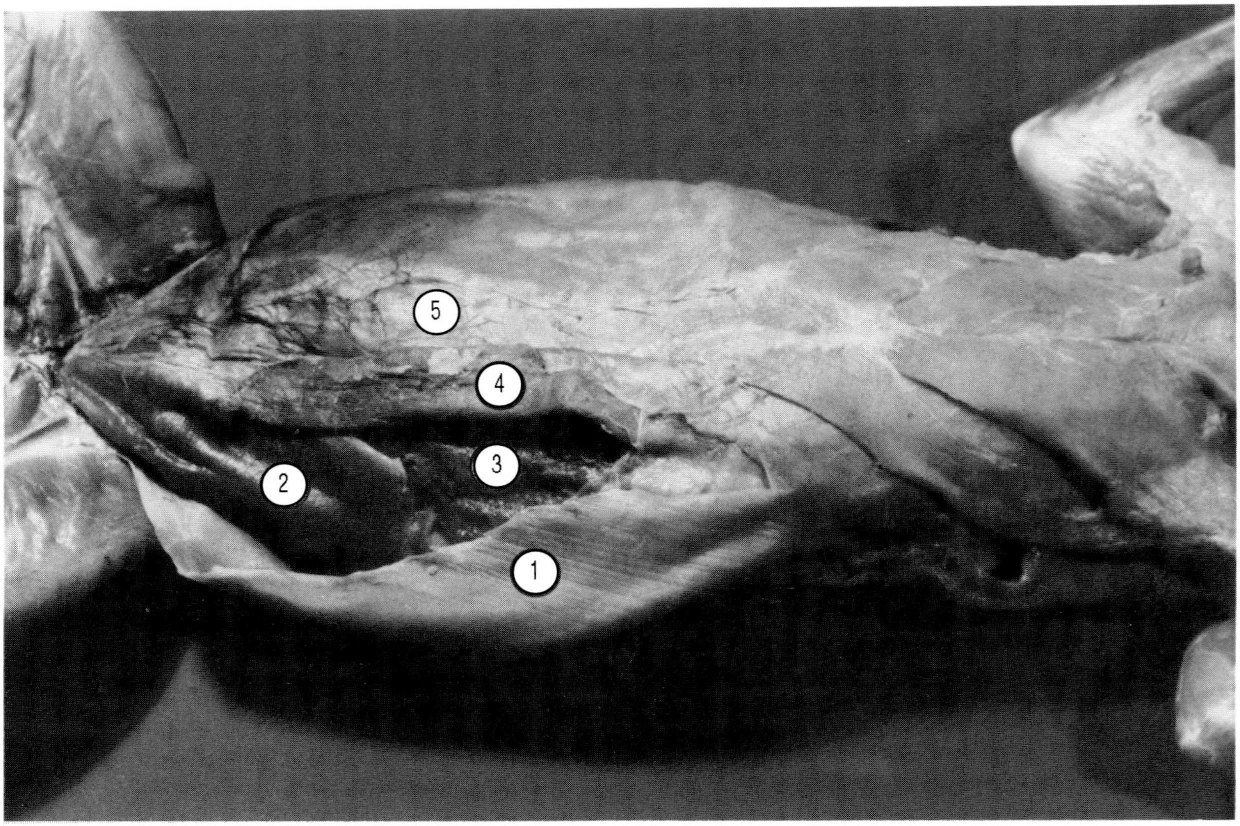

1. **External abdominal oblique**
2. **Internal abdominal oblique**
   - **A:** Compresses abdomen
   - **O:** Pelvis, iliac crest; L_4 - L_7 vertebrae
   - **I:** By aponeurosis on the linea alba
3. **Transversus abdominis**
4. **Rectus** (REK-tus) **abdominis**
5. **Aponeurosis of the external abdominal oblique** overlying the right rectus abdominis muscle

FIGURE 7 SUPERFICIAL MUSCLES OF THE NECK AND THROAT - VENTROLATERAL VIEW (RIGHT SIDE)

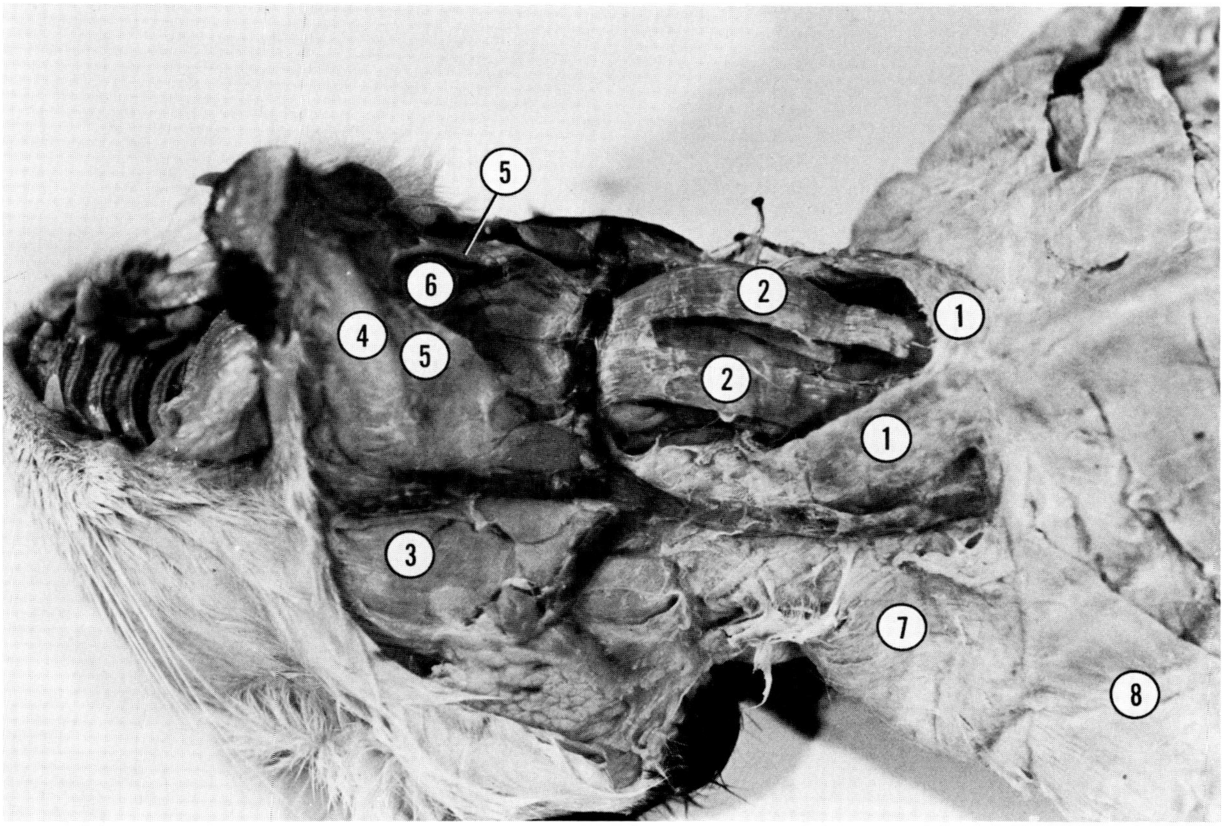

1. **Sternomastoid** (stern-nō-MAS-toyd)
   - **A:** Rotates and lowers head
   - **O:** Sternum, manubrium portion
   - **I:** Skull, mastoid process
2. **Sternohyoid** (ster´ nō-HĪ-oyd)
   - **A:** Pulls hyoid caudad
   - **O:** Manubrium & 1st costal cartilage
   - **I:** Hyoid
3. **Masseter** (ma-SE-ter)
   - **A:** Elevates mandible
   - **O:** Zygomatic arch
   - **I:** Mandible, coronoid fossa
4. **Mandible** (MAN-di-bal)
5. **Digastric** (dī-GAS-trik)
   - **A:** Depresses mandible
   - **O:** Skull, mastoid process
   - **I:** Mandible
6. **Mylohyoid** (mī´ lō-HĪ-oyd)
   - **A:** Elevates the floor of the mouth
   - **O:** Larynx, cricoid cartilage
   - **I:** Mid-ventral raphe
7. **Clavotrapezius** (klav´ ō-tra-PĒ-zē-us)
   - **A:** Moves scapula medially
   - **O:** Superior nuchal line
   - **I:** Scapula, spine and metacromion process
8. **Clavodeltoid** (klav´ ō-DEL-toyd) or **Clavobrachialis** (klav´ ō-brā-kē-Ā-lis)

FIGURE **8** SUPERFICIAL MUSCLES OF THE NECK AND THROAT - VENTRAL VIEW

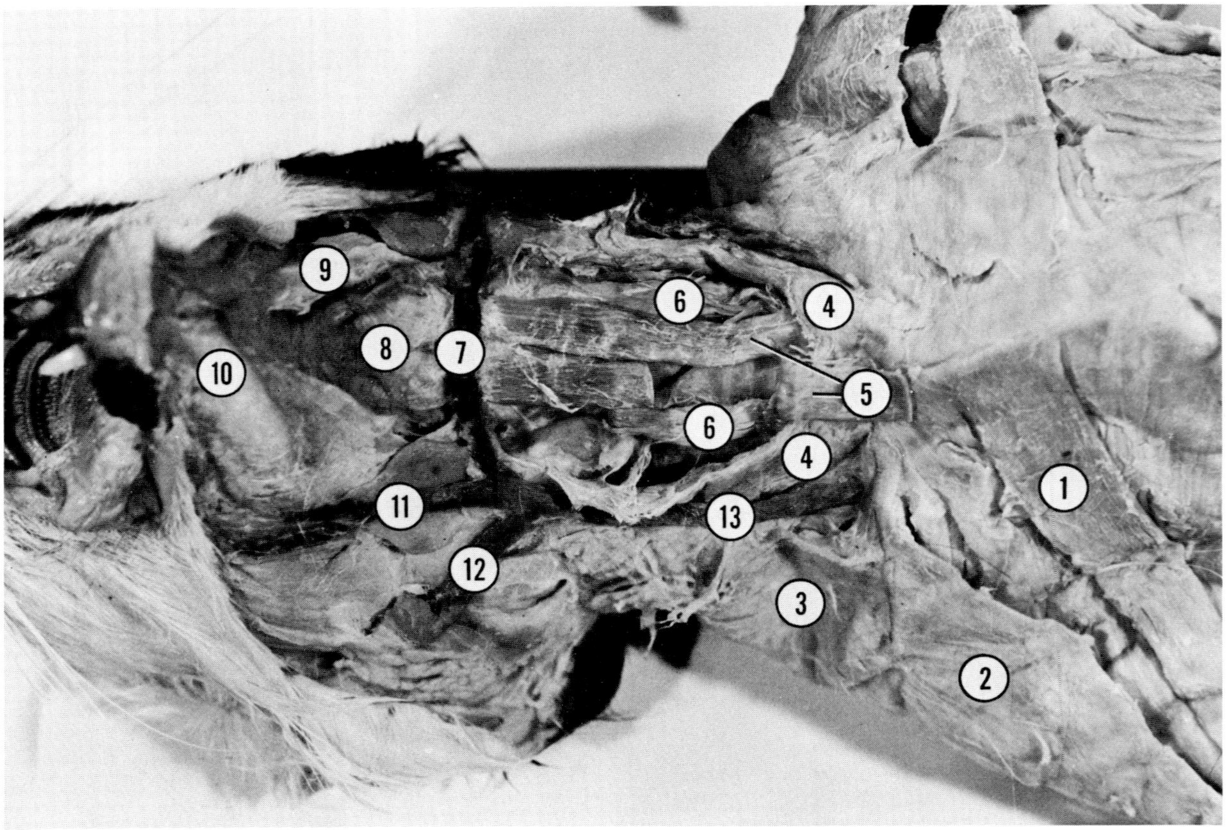

1. **Pectoantibrachialis** (pek´ tō-an-tē-brā-kē-Ā-lis)
2. **Clavodeltoid (Clavobrachialis)**
3. **Clavotrapezius**
4. **Sternomastoid**
5. **Sternohyoid**
6. **Sternothyroid** (ster´ nō-THĪ-royd)
   - **A:** Pulls larynx caudad
   - **O:** Beneath sternohyoid
   - **I:** Thyroid cartilage, posterior border
7. **Transverse jugular** (JUG-yoo-lar) **vein**
8. **Mylohyoid**
9. **Digastric**
10. **Mandible**
11. **Anterior facial vein**
12. **Posterior facial vein**
13. **External jugular vein**

FIGURE 9 SUPERFICIAL MUSCLES, GLANDS, AND BLOOD VESSELS OF THE NECK AND HEAD - LEFT LATERAL VIEW

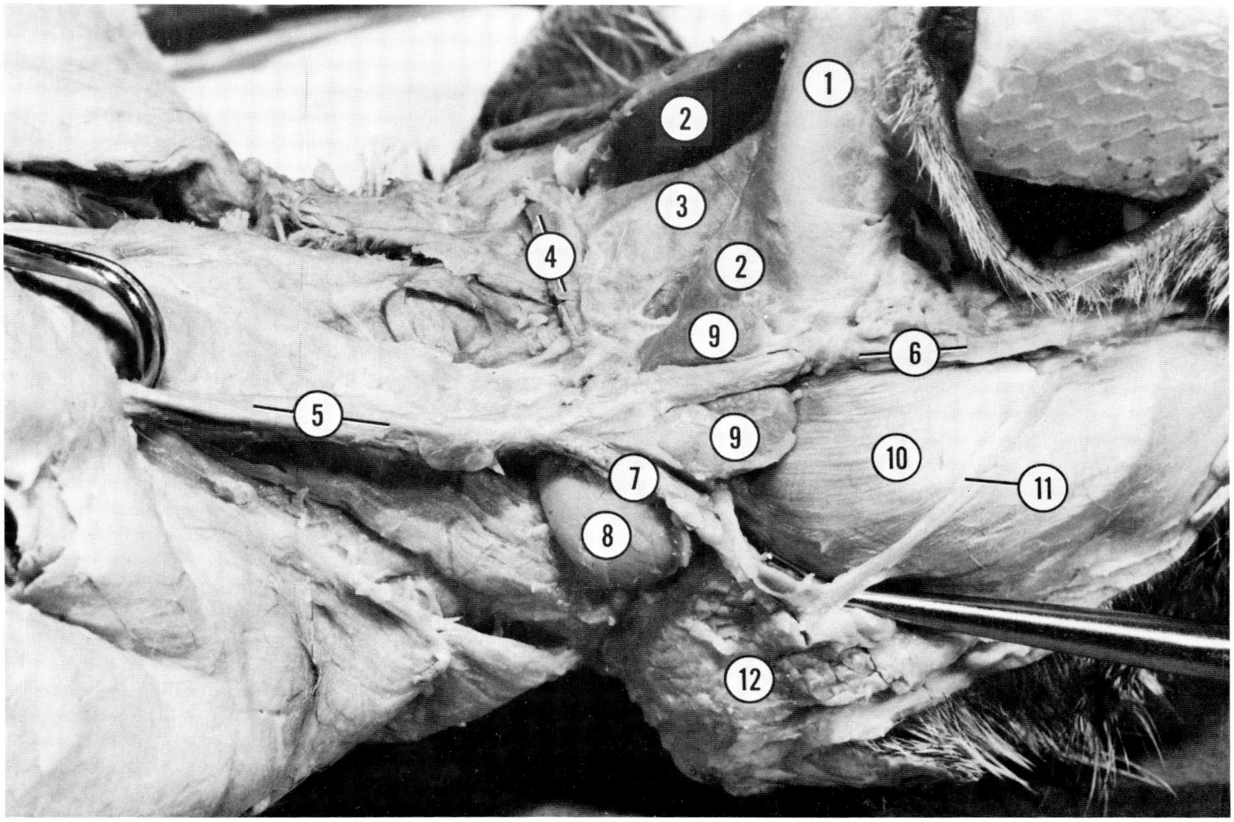

1. **Mandible**
2. **Digastric**
3. **Mylohyoid**
4. **Transverse jugular vein**
5. **Left external jugular vein**
6. **Left anterior facial vein**
7. **Left posterior facial vein**
8. **Submandibular** (sub´ man-DIB-yoo-lar) **gland**
9. **Lymph** (limf) **nodes**
10. **Masseter muscle**
11. **Parotid** (pa-ROT-id) **duct**
12. **Parotid gland**

FIGURE 10 DEEP MUSCLES OF THE NECK AND THROAT REGION - VENTRAL VIEW

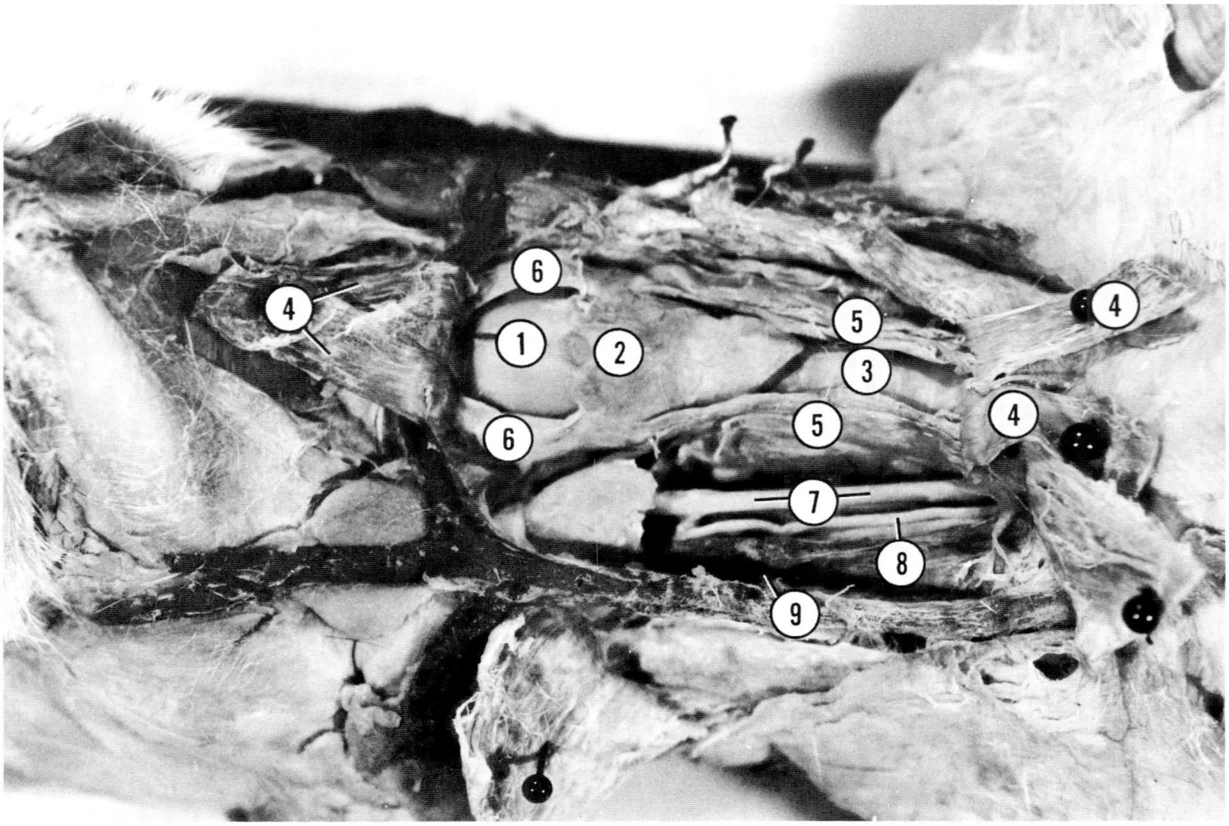

1. **Thyroid cartilage of larynx** (THĪ-royd kar-ti-lij LAR-inks)
2. **Region of cricoid** (KRĪ-koyd) **cartilage of larynx**
3. **Trachea** (TRĀ-kē-a)
4. **Sternohyoid muscles,** transected and reflected
5. **Sternothyroid**
6. **Thyrohyoid** (thī-rō-HĪ-oyd)
   - **A:** Pulls hyoid bone caudad and dorsad
   - **O:** Thyroid cartilage, lateral surface
   - **I:** Thyrohyoid bone, medial surface
7. **Right common carotid** (ka-ROT-id) **artery**
8. **Right vagus** (VĀ-gus) **nerve**
9. **Right external jugular vein**

FIGURE **11** SUPERFICIAL TRUNK MUSCLES - DORSOLATERAL VIEW (LEFT SIDE)

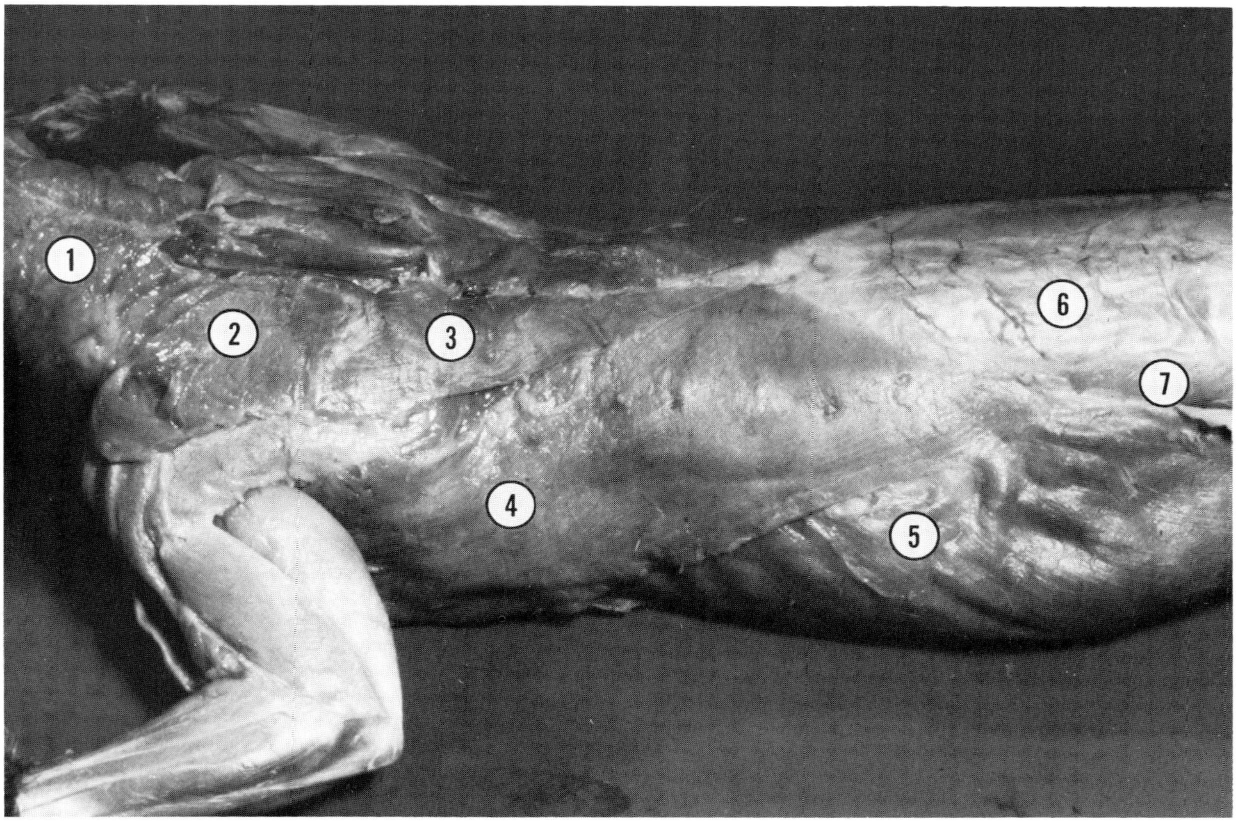

1. **Clavotrapezius** (klav´ ō-tra-PĒ-zē-us)
2. **Acromiotrapezius** (a-krō-mē-ō-tra-PĒ-zē-us)
3. **Spinotrapezius** (spī-nō-tra-PĒ-zē-us)
4. **Latissimus dorsi**
5. **External abdominal oblique**
6. **Superficial lumbodorsal fascia** (lum-bō-DOR-sul FASH-a)
7. **Longissimus** (lon-JIS-i-mus) **dorsi**

FIGURE 12 SUPERFICIAL MUSCLES OF THE SHOULDER - LEFT LATERAL VIEW

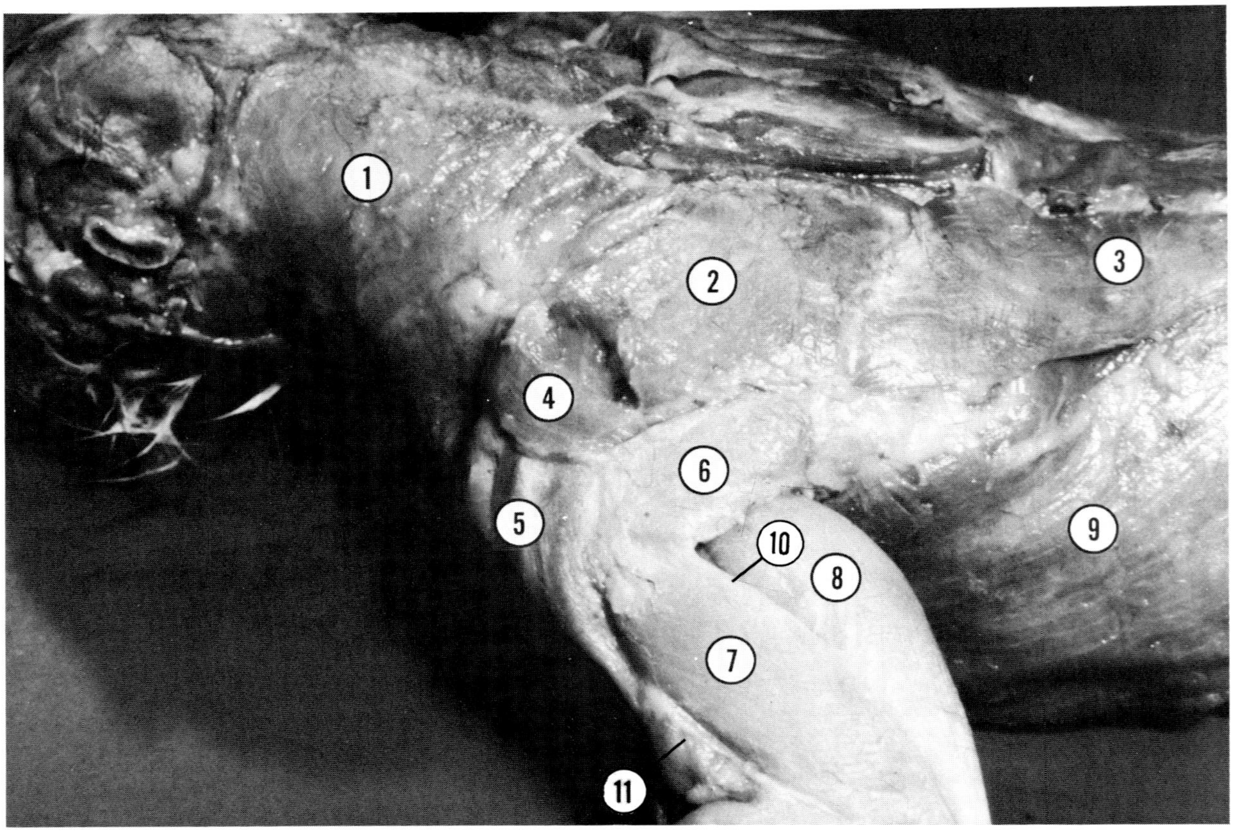

1. **Clavotrapezius** (klav´ ō-tra-PĒ-zē-us)
2. **Acromiotrapezius** (a-krō-mē-ō-tra-PĒ-zē-us)
   - **A:** Holds scapulae together
   - **O:** Spinous process of T_1 & T_4 vertebrae
   - **I:** Scapula, scapular spine
3. **Spinotrapezius** (spī-nō-tra-PĒ-zē-us)
   - **A:** Pulls scapula caudad and dorsad
   - **O:** Thoracic vertebrae, spinous process
   - **I:** Scapula, spine; into fascia of supraspinatus & infraspinatus m.
4. **Levator scapulae ventralis** (le-VĀ-ter SKAP-yoo-lā ven-TRAL-is)
   - **A:** Pulls scapula craniad
   - **O:** Occipital bone, ventral surface; C_1 vertebra
   - **I:** Scapula, metacromion and infraspinatus fossa
5. **Acromiodeltoid** (a-krō-mē-ō-DEL-toyd)
   - **A:** Flexes and rotates humerus
   - **O:** Scapula, posterior border of acromion
   - **I:** Spinodeltoid muscle
6. **Spinodeltoid** (spī-nō-DEL-toyd)
   - **A:** Flexes and outwardly rotates humerus
   - **O:** Scapula, spine
   - **I:** Humerus, deltoid ridge
7. **Triceps brachii lateralis** (TRĪ-seps BRĀ-kē-ī lat-er-AL-is))
   - **A:** Extends elbow
   - **O:** Humerus
   - **I:** Ulna, olecranon process
8. **Triceps brachii, longus**
   - **A:** Extends elbow
   - **O:** Scapula, posterior segment
   - **I:** Ulna, olecranon process
9. **Latissimus dorsi** (la-TIS-i-mus DOR-sē)
10. **Triceps brachii, medial head** (not visible)
    - **A:** Extends elbow
    - **O:** Humerus
    - **I:** Ulna, olecranon process
11. **Clavodeltoid** or **Clavobrachialis**

FIGURE 13 DEEP MUSCLES OF THE SHOULDER AND SCAPULA -DORSOLATERAL VIEW

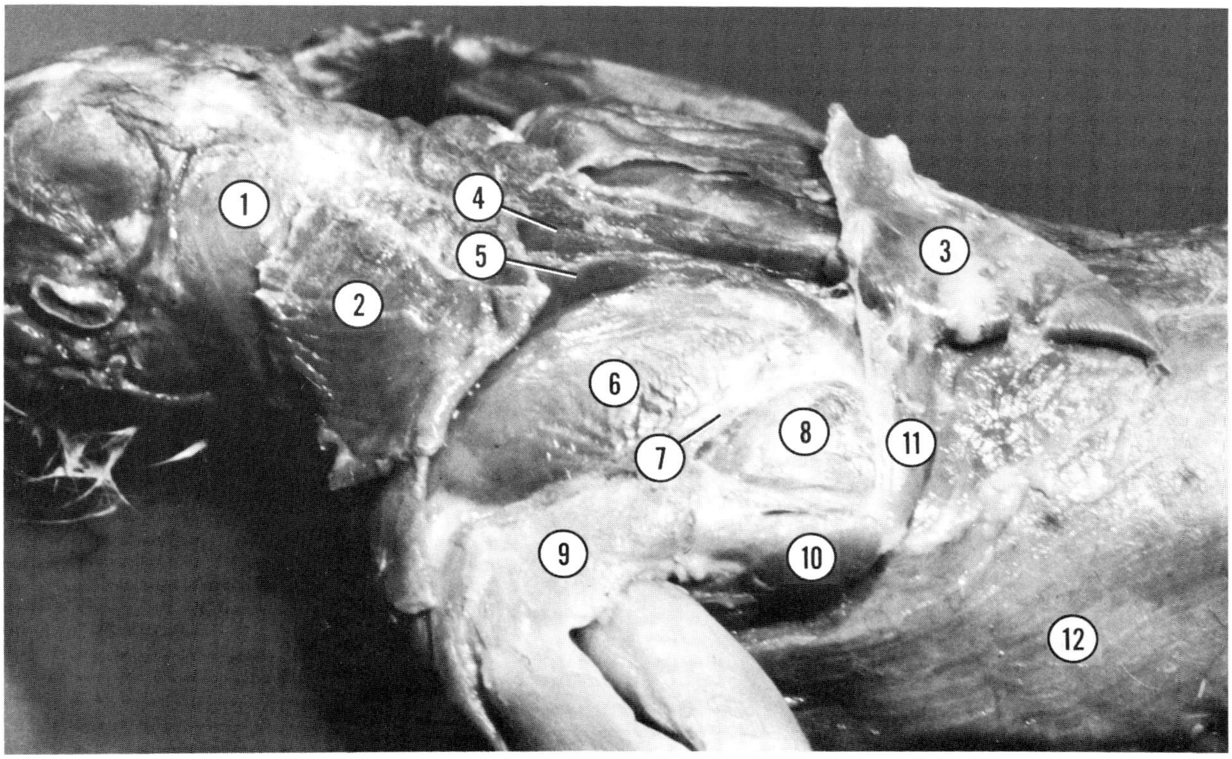

1. **Clavotrapezius**
2. **Acromiotrapezius,** transected and reflected
3. **Spinotrapezius,** transected and reflected
4. **Rhomboideus** (rom-BOY-dē-us) **minor**
   - **A:** Pulls scapula craniad
   - **O:** Spinous processes of C_1 - C_7 & T_1 - T_4 vertebrae
   - **I:** Scapula, vertebral border
5. **Rhomboideus capitis** (KAP-i-tis) or **Occipitoscapularis** (ok-sip´ i-tō-skap-yoo-LAR-is)
   - **A:** Rotates and draws forward the scapula
   - **O:** Occipital, superior nuchal line
   - **I:** Scapula, vertebral border
6. **Supraspinatus** (soo pra-spī-NĀ-tus)
   - **A:** Extension of humerus
   - **O:** Scapula, supraspinatus fossa
   - **I:** Shoulder joint and the greater tubercle
7. **Scapular** (SKAP-yoo-lar) **spine**
8. **Infraspinatus** (in´ fra-spī-NĀ-tus)
   - **A:** Abduction and outward rotation of humerus
   - **O:** Scapula, infraspinatus fossa
   - **I:** Humerus, greater tuberosity
9. **Spinodeltoid**
10. **Teres** (TER-ez) **major**
    - **A:** Inwardly rotates the humerus
    - **O:** Posterior border of scapula
    - **I:** Humerus, shaft
11. **Rhomboideus major** (insertion end)
    - **A:** Pulls scapula caudad & dorsad
    - **O:** Spinous processes of C_1 - C_7 & T_1 - T_4 vertebrae
    - **I:** Scapula, vertebral border
12. **Latissimus dorsi**

FIGURE 14 DEEP MUSCLES OF THE SCAPULA - DORSOMEDIAL VIEW

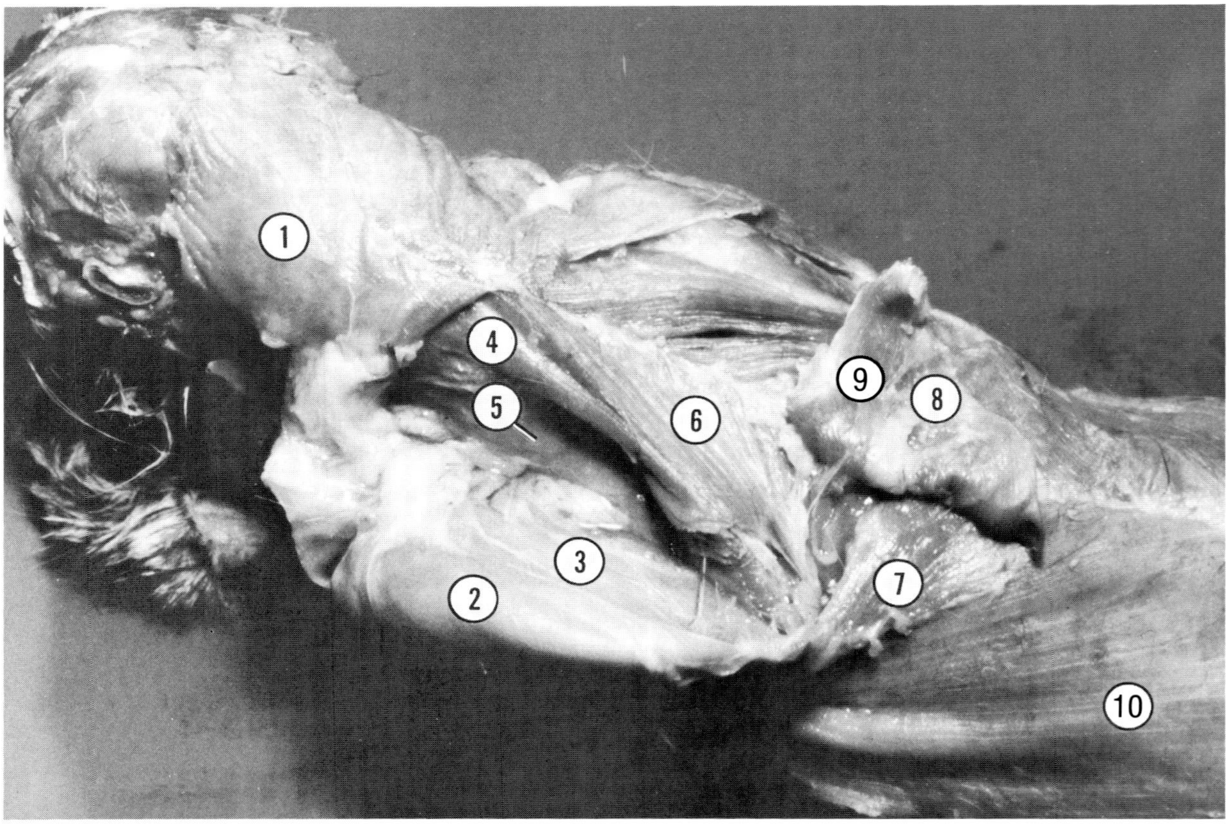

1. **Clavotrapezius**
2. **Supraspinatus**
3. **Subscapularis** (sub-skap´ yoo-LAR-is)
   - **A:** Adduction and extension of shoulder joint
   - **O:** Scapula, subscapular fossa
   - **I:** Humerus, posterior border
4. **Rhomboideus capitis**
5. **Splenius** (SPLĒ-nē-us)
   - **A:** Elevation and lateral flexion of head
   - **O:** T_1 vertebra, thoracic spine
   - **I:** Occipital, lambdoidal ridge
6. **Rhomboideus minor**
7. **Rhomboideus major**
8. **Spinotrapezius,** transected and reflected
9. **Acromiotrapezius,** transected and reflected
10. **Latissimus dorsi**

FIGURE **15** MUSCLES OF THE UPPER FORELIMB - LEFT MEDIAL VIEW

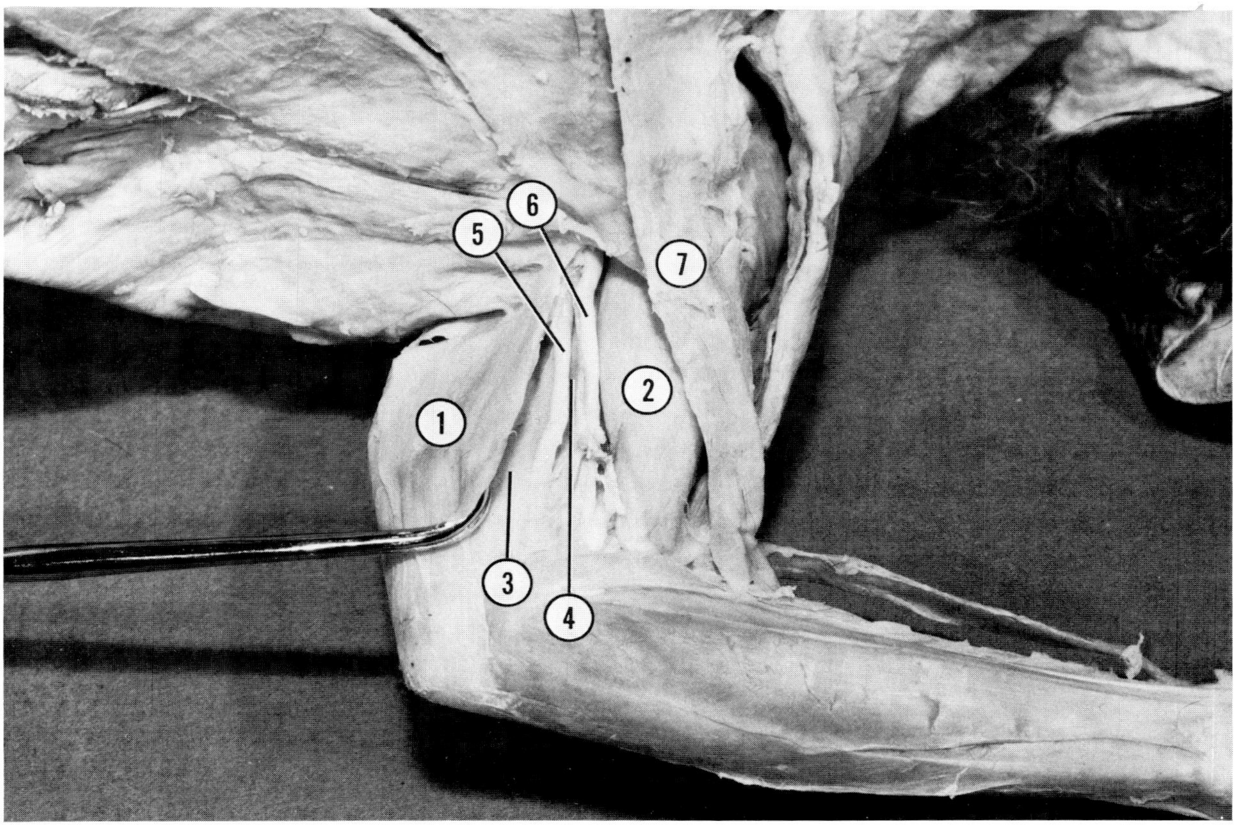

1. **Epitrochlearis** (ep´ ē-trok´ lē-Ā-ris)
   - **A:** Extension of antibrachium
   - **O:** Latissimus dorsi, inferior border
   - **I:** Ulna, olecranon process
2. **Biceps brachii** (BĪ-ceps BRĀ-kē-ī)
   - **A:** Flexion of forearm; supination of hand
   - **O:** Scapula, glenoid tubercle
   - **I:** Radius, radial tuberosity
3. **Triceps brachii, longus**
4. **Left brachial** (BRĀ-kē-al) **artery**
5. **Ulnar** (UL-nar) **nerve**
6. **Median** (MĒ-dē-an) **nerve**
7. **Pectoantibrachialis** (pek´ tō-an´ tē-brā´ kē-Ā-lis)

FIGURE 16 MUSCLES OF THE LEFT FORELIMB PRIOR TO DISSECTION - MEDIAL VIEW

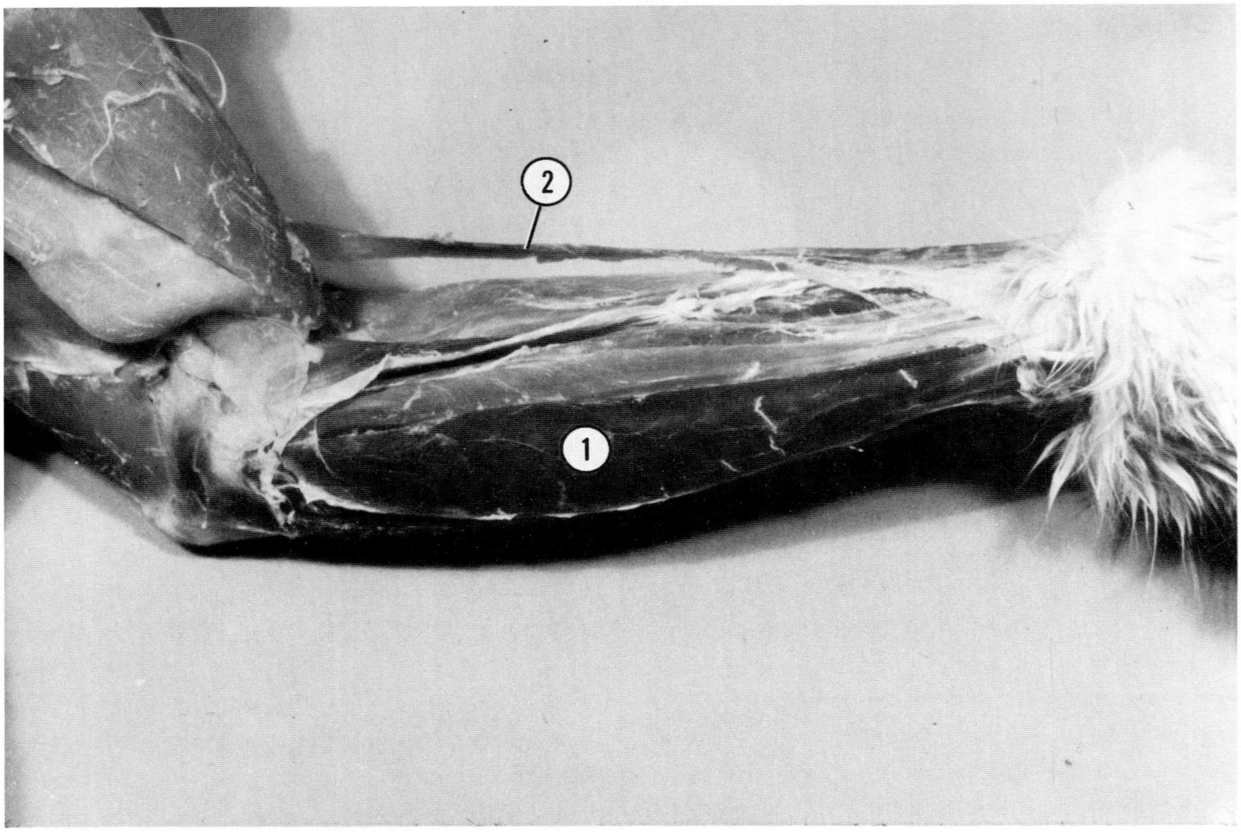

1. **Palmaris** (pal-MĀ-ris) **longus**
   - **A:** Digit flexor first phalanx
   - **O:** Humerus, medial epicondyle
   - **I:** Proximal phalanx, base

2. **Brachioradialis** (brā´ kē-ō-rā-dē-Ā-lis)
   - **A:** Supination of hand
   - **O:** Humerus, posterior border
   - **I:** Radius, styloid process

FIGURE 17 MUSCLES OF THE FORELIMB - LEFT MEDIAL VIEW

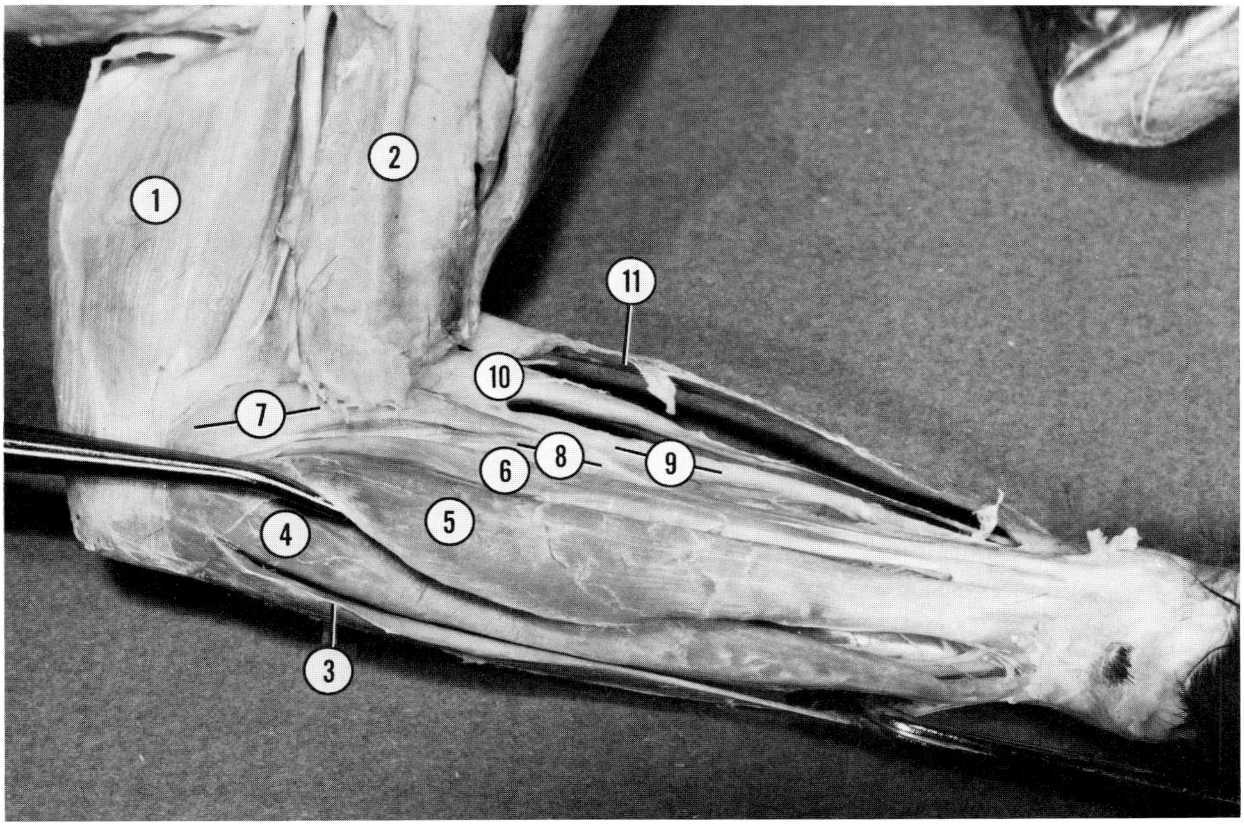

1. **Epitrochlearis**
2. **Pectoantibrachialis** (pek´ tō-an´ tē-brā-kē-Ā-lis)
3. **Flexor carpi ulnaris, first head** (FLEK-ser KAR-pī UL-na-ris)
   - **A:** Flexes wrist
   - **O:** Humerus, distal end
   - **I:** Pisiform bone
4. **Flexor carpi ulnaris, second head**
   - **A:** Flexes wrist
   - **O:** Ulnar, olecranon process
   - **I:** Pisiform bone
5. **Palmaris** (pal-MĀ-ris) **longus**
6. **Flexor carpi radialis** (rā-dē-Ā-lis)
   - **A:** Flexes wrist
   - **O:** Humerus, medial epicondyle
   - **I:** Base of metacarpals, 2 and 3
7. **Pronator teres** (prō-NĀ-ter TER-ez)
   - **A:** Pronation of hand
   - **O:** Humerus, medial epicondyle
   - **I:** Radius, medial border
8. **Flexor digitorum profundus, third head** (di-ji-TOR-um prō-FUN-dus)
   - **A:** Flexes all digits
   - **O:** Humerus, radius and ulna
   - **I:** Base of phalanges, distal
9. **Extensor carpi radialis brevis** (BREV-is)
   - **A:** Extends hand
   - **O:** Humerus, lateral epicondyle
   - **I:** Third, metacarpal, base
10. **Extensor carpi radialis longus**
    - **A:** Extends hand
    - **O:** Humerus, lateral epicondyle
    - **I:** Second metacarpal, base
11. **Brachioradialis**

FIGURE 18 MUSCLES OF THE FORELIMB - LEFT LATERAL VIEW

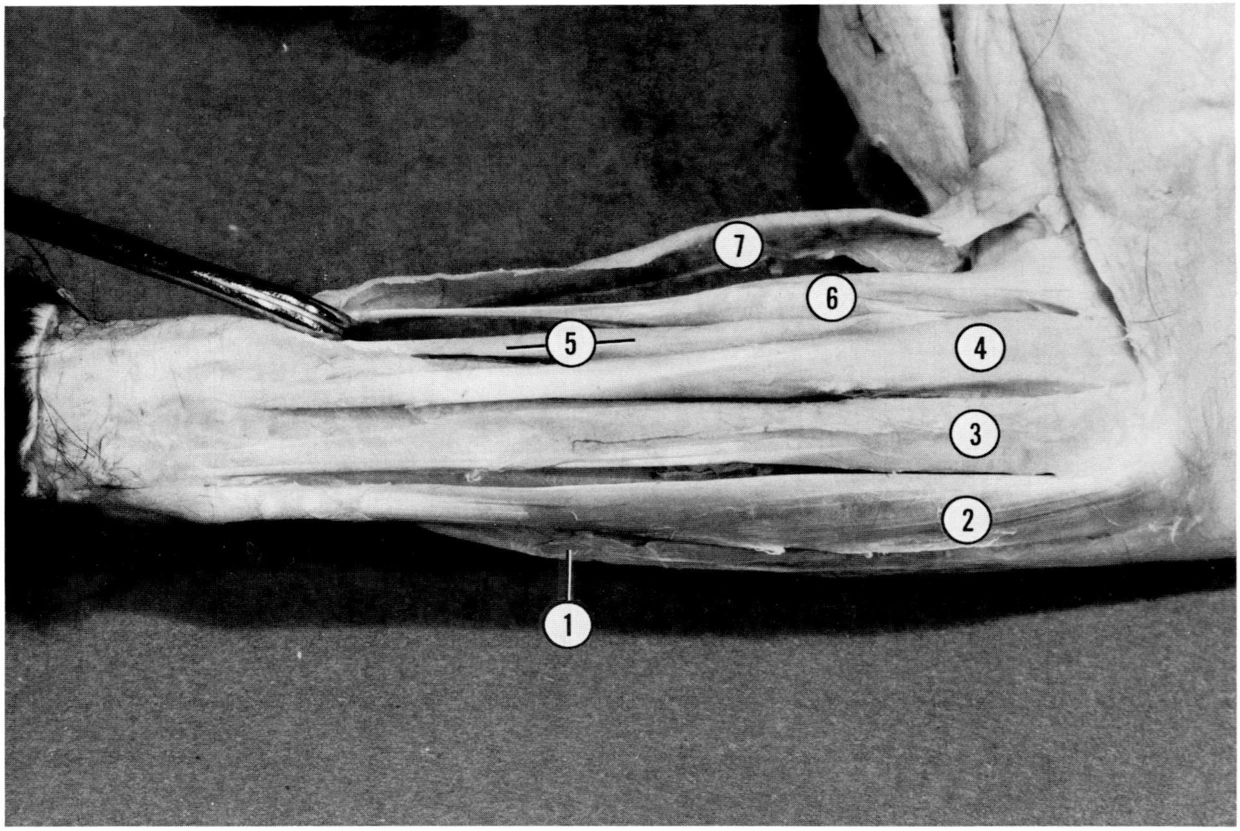

1. **Flexor carpi ulnaris, first head**
2. **Extensor carpi ulnaris**
   - **A:** Extends carpal joint
   - **O:** Humerus, lateral epicondyle
   - **I:** Fifth metacarpal, ulnar side
3. **Extensor digitorum lateralis** (di-ji-TOR-um lat ´er-AL-is)
   - **A:** Extends digits
   - **O:** Humerus, supracondyloid ridge
   - **I:** Common tendons on digits 3 to 5
4. **Extensor digitorum communis** (kom-YOO-nis)
   - **A:** Extends digits
   - **O:** Humerus, lateral epicondyle
   - **I:** Second phalanx, base
5. **Extensor carpi radialis, short head**
6. **Extensor carpi radialis, long head**
   - **A:** Extends hand
   - **O:** Humerus, lateral epicondyle
   - **I:** Base of metacarpals, 2 and 3
7. **Brachioradialis**

FIGURE 19 SUPERFICIAL MUSCLES OF THE THIGH REGION -VENTROMEDIAL VIEW

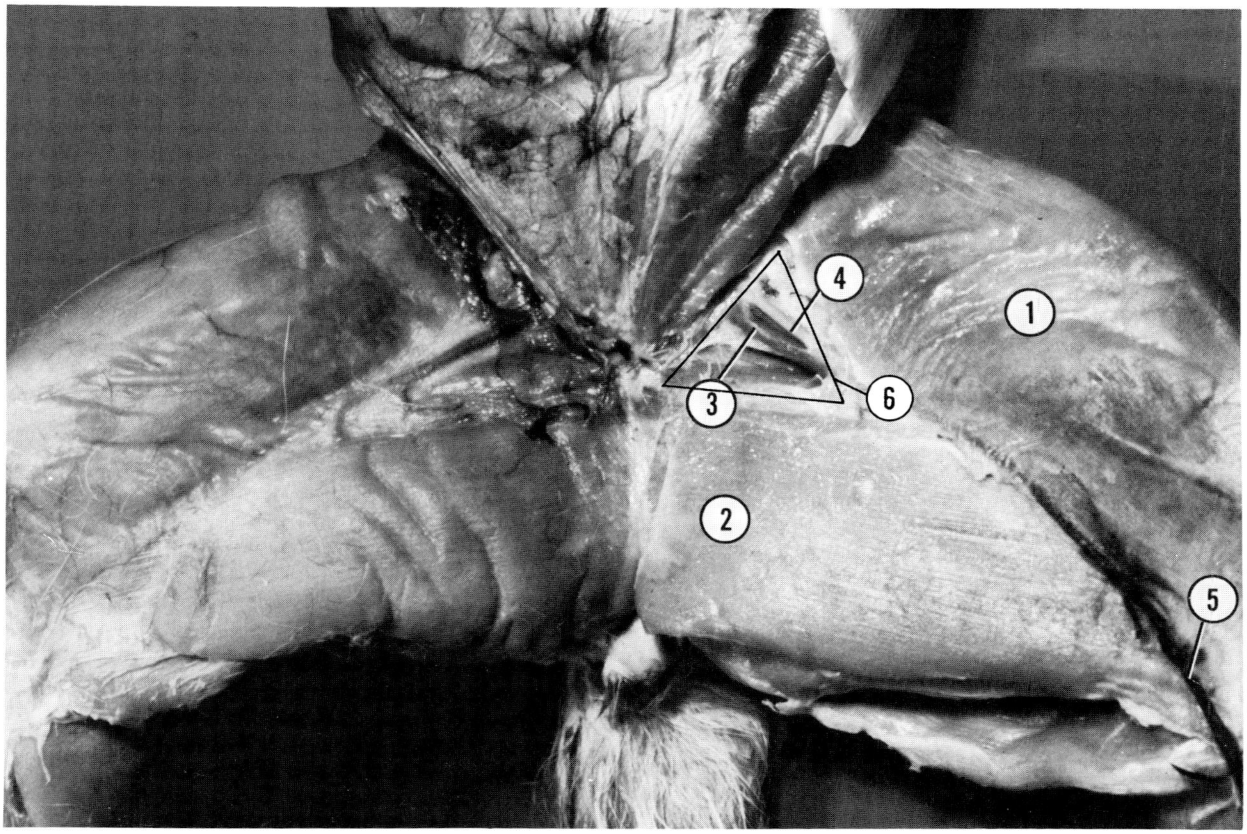

1. **Sartorius** (sar-TOR-ē-us)
   - **A:** Flexes thigh and shank
   - **O:** Ilium, crest
   - **I:** Tibia, shaft-proximal end; patella
2. **Gracilis** (GRAS-i-lis)
   - **A:** Adductor and extensor of leg
   - **O:** Pubis, pubic symphysis - posterior three-fourths
   - **I:** Tibia, proximal end - medial surface
3. **Left femoral** (FEM-or-ul) **vein**
4. **Left femoral artery**
5. **Left saphenous** (SAF-e-nus) **vein**
6. **Femoral triangle**

FIGURE 20 DEEP MUSCLES OF THE LEFT THIGH - VENTROMEDIAL VIEW (SARTORIUS AND GRACILIS MUSCLES TRANSECTED AND REFLECTED)

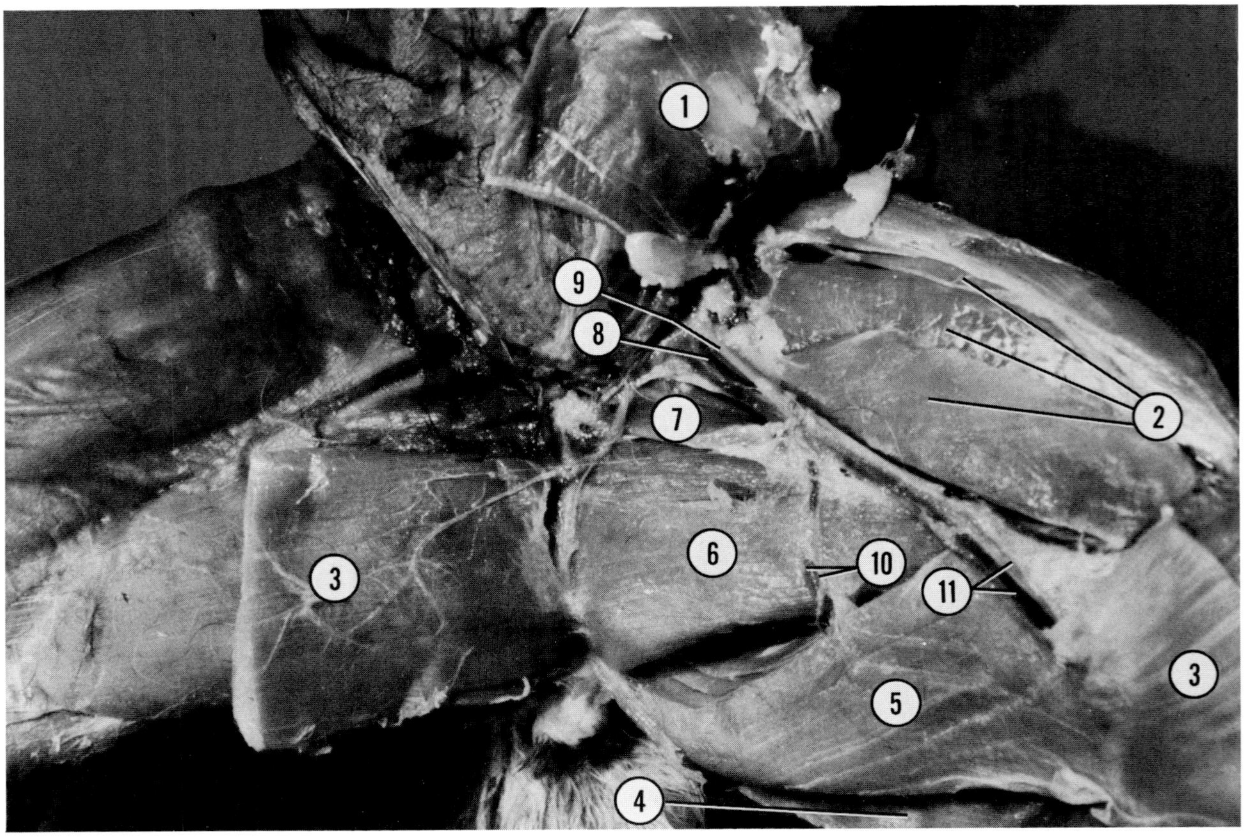

1. **Sartorius** (sar-TOR-ē-us), transected and reflected
2. **Quadriceps** (KWOD-ri-seps) **group**
   - **A:** Extends shank
   - **O:** Femur, shaft; ilium, ventral border
   - **I:** Patella, medial and proximal border
3. **Gracilis,** transected and reflected
4. **Semitendinosus** (sem´ ē-ten´ di-NŌ-sus)
   - **A:** Extends thigh; flexes shank
   - **O:** Ischium
   - **I:** Tibia, proximal end
5. **Semimembranosus** (sem´ ē-mem´ bra-NŌ-sus)
   - **A:** Flexes shank; extends thigh
   - **O:** Ischium, ramus
   - **I:** Femur, medial surface
6. **Adductor femoris** (fem-OR-is)
   - **A:** Adducts and extends thigh
   - **O:** Pubis, pubic symphysis
   - **I:** Femur, posterior surface
7. **Adductor longus**
   - **A:** Adducts and flexes thigh
   - **O:** Pubis
   - **I:** Femur, linea aspera
8. **Left femoral vein**
9. **Left femoral artery**
10. **Muscular branch of femoral artery and vein**
11. **Left saphenous** (SAF-e-nus) **artery and vein**

FIGURE 21 DEEP MUSCLES OF THE LEFT THIGH SHOWING QUADRICEPS MUSCLE DETAIL - MEDIAL VIEW

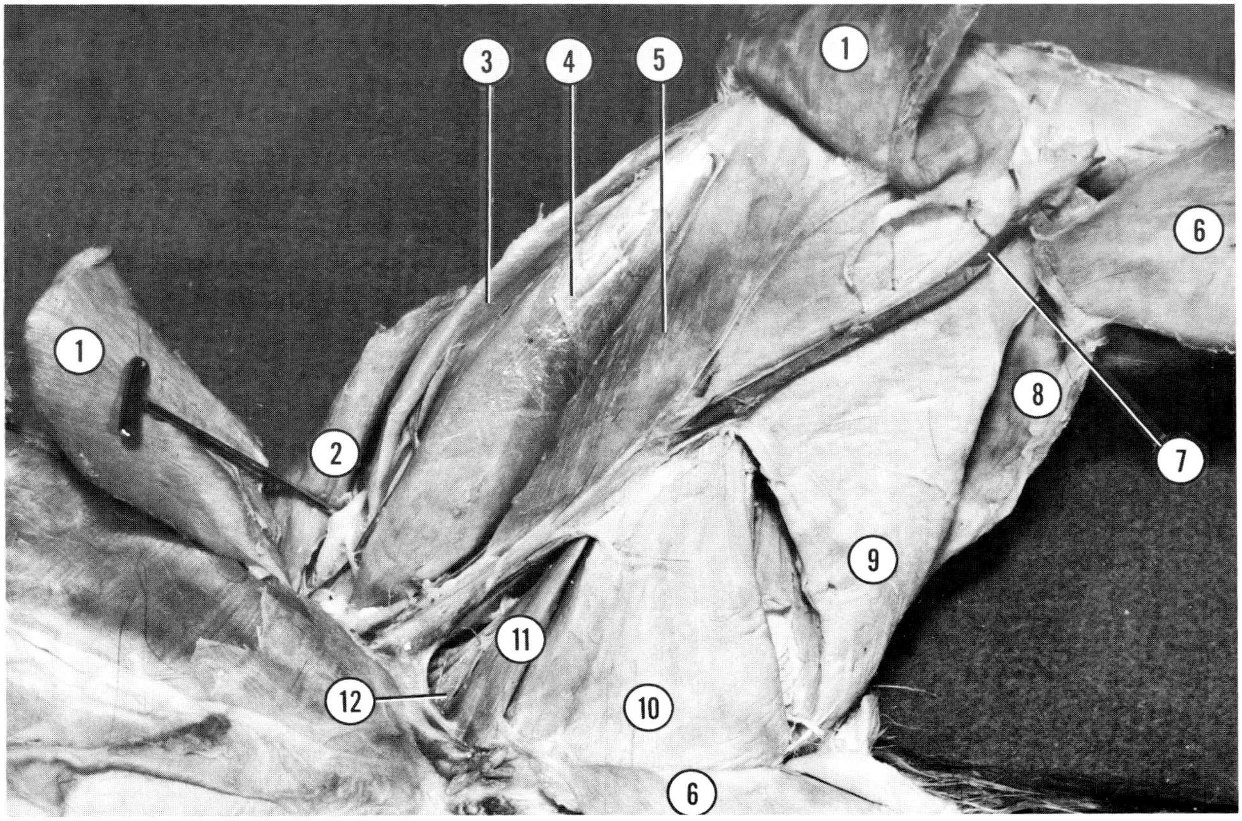

1. **Sartorius,** transected & reflected
2. **Tensor fasciae latae** (fa´ SHĀ LĀ-tā)
   - **A:** Extends leg and flexes hip joint
   - **O:** Ilium, inferior border
   - **I:** Fascia latae
3. **Vastus lateralis** (VAS-tus lat´ er-AL-is)
   - **A:** Extends shank
   - **O:** Femur, shaft and greater trochanter
   - **I:** Patella, outer surface
4. **Rectus femoris** (REK-tus fem-OR-is)
   - **A:** Extends shank
   - **O:** Acetabulum; ileum, inferior border
   - **I:** Patella, outer surface
5. **Vastus medialis** (mē´ dē-AL-is)
   - **A:** Extends shank
   - **O:** Femur, shaft
   - **I:** Patella, medial border
6. **Gracilis,** transected and reflected
7. **Saphenous artery and vein**
8. **Semitendinosus**
9. **Semimembranosus**
10. **Adductor femoris**
11. **Adductor longus**
12. **Pectineus** (pek-TIN-ē-us)
    - **A:** Adducts thigh
    - **O:** Pubis, anterior border
    - **I:** Femur, shaft

FIGURE **22** SUPERFICIAL MUSCLES OF THE LEFT HINDLIMB - LATERAL VIEW

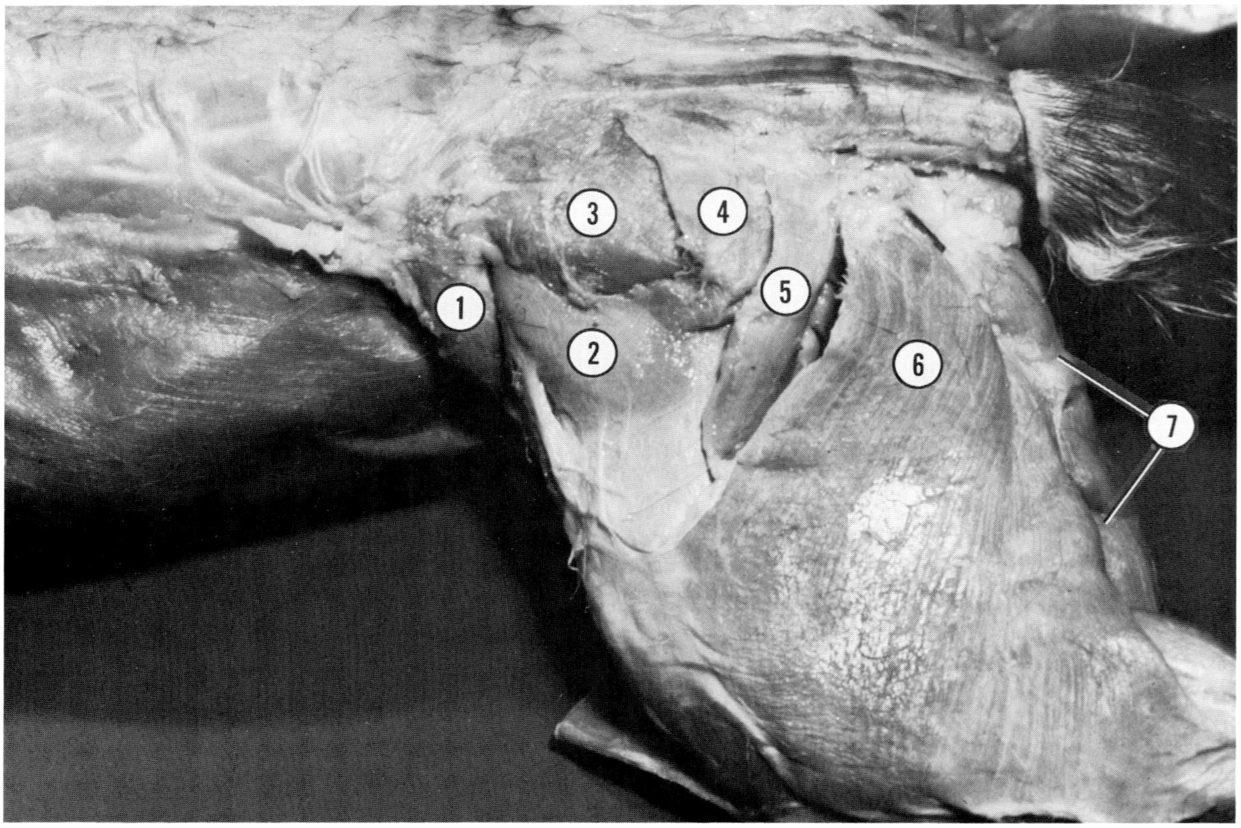

1. **Sartorius**
2. **Tensor fasciae latae**
3. **Gluteus medius** (GLOO-tē-us MĒ-dē-us)
   - **A:** Abduction and extension of thigh
   - **O:** Ilium, lateral border; S_7 and C_2 vertebrae
   - **I:** Femur, greater trochanter
4. **Gluteus maximus** (MAK-si-mus)
   - **A:** Abduction and extension of thigh
   - **O:** S_7 and C_1 vertebrae, transverse processes
   - **I:** Femur, greater trochanter
5. **Caudofemoralis** (kaw-dō-fem´ or-AL-is)
   - **A:** Abduction of thigh; extension of lower leg
   - **O:** Caudal vertebrae, transverse process
   - **I:** Patella, lateral border
6. **Biceps femoris** (BĪ-seps fem-OR-is)
   - **A:** Abduction and extension of thigh; flexes lower leg
   - **O:** Ischium, posterior surface
   - **I:** Tibia, posterior border; patella, lateral margin
7. **Semitendinosus** (sem´ ē-ten´di-NŌ-sus)

FIGURE 23 DEEP MUSCLES OF THE HINDLIMB - LEFT LATERAL VIEW

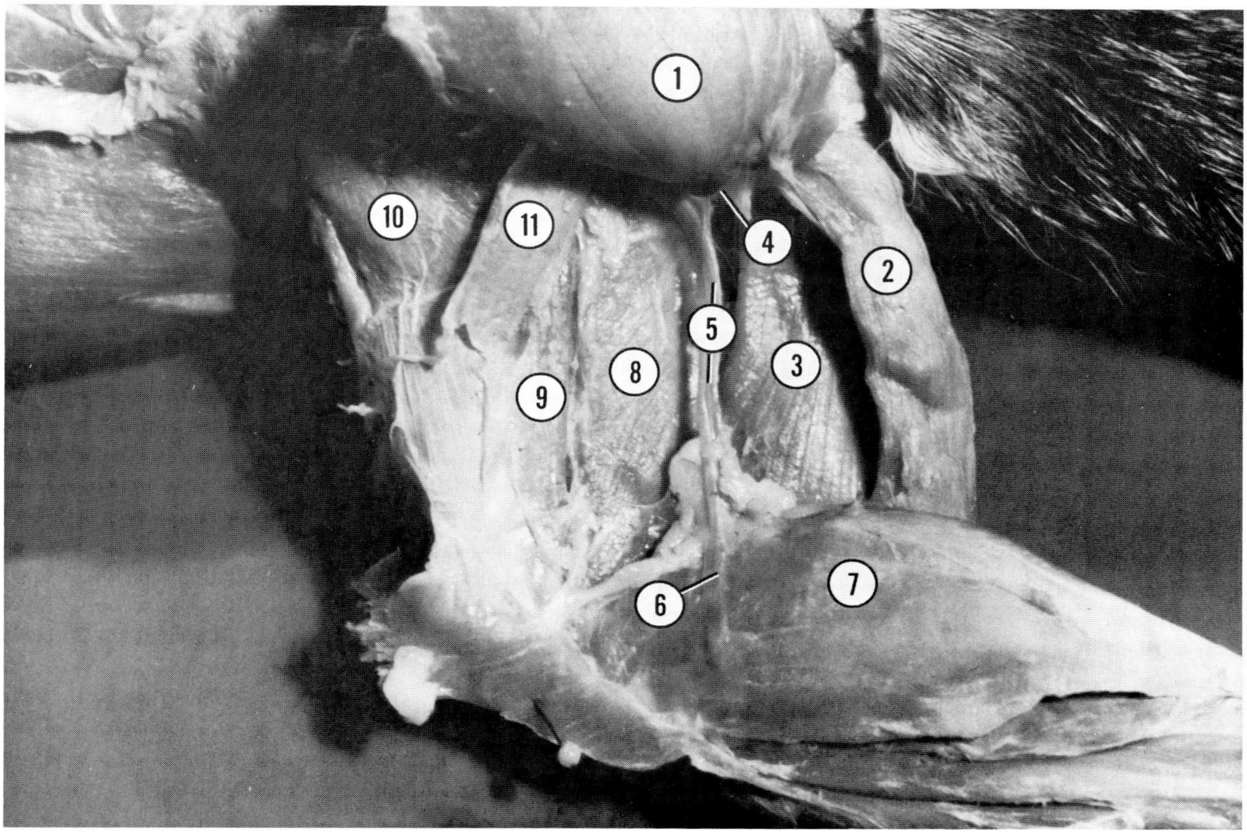

1. **Biceps femoris,** reflected
2. **Semitendinosus**
3. **Semimembranosus**
4. **Sciatic** (sī-AT-ik) **nerve, muscular branch**
5. **Sciatic nerve**
6. **Common peroneal** (per´ ō-NĒ-ul) **nerve**
7. **Gastrocnemius** (gas´ trok-NĒ-mē-us) or (gas´ trō-NĒ-mē-us)
   - **A:** Extends foot and flexes shank
   - **O:** Femur, distal end - posterior surface
   - **I:** Calcaneus
8. **Adductor femoris**
9. **Vastus lateralis**
10. **Tensor fasciae latae**
11. **Caudofemoralis**

FIGURE 24 MUSCLES AND NERVES OF THE LEFT HINDLIMB - LATERAL VIEW

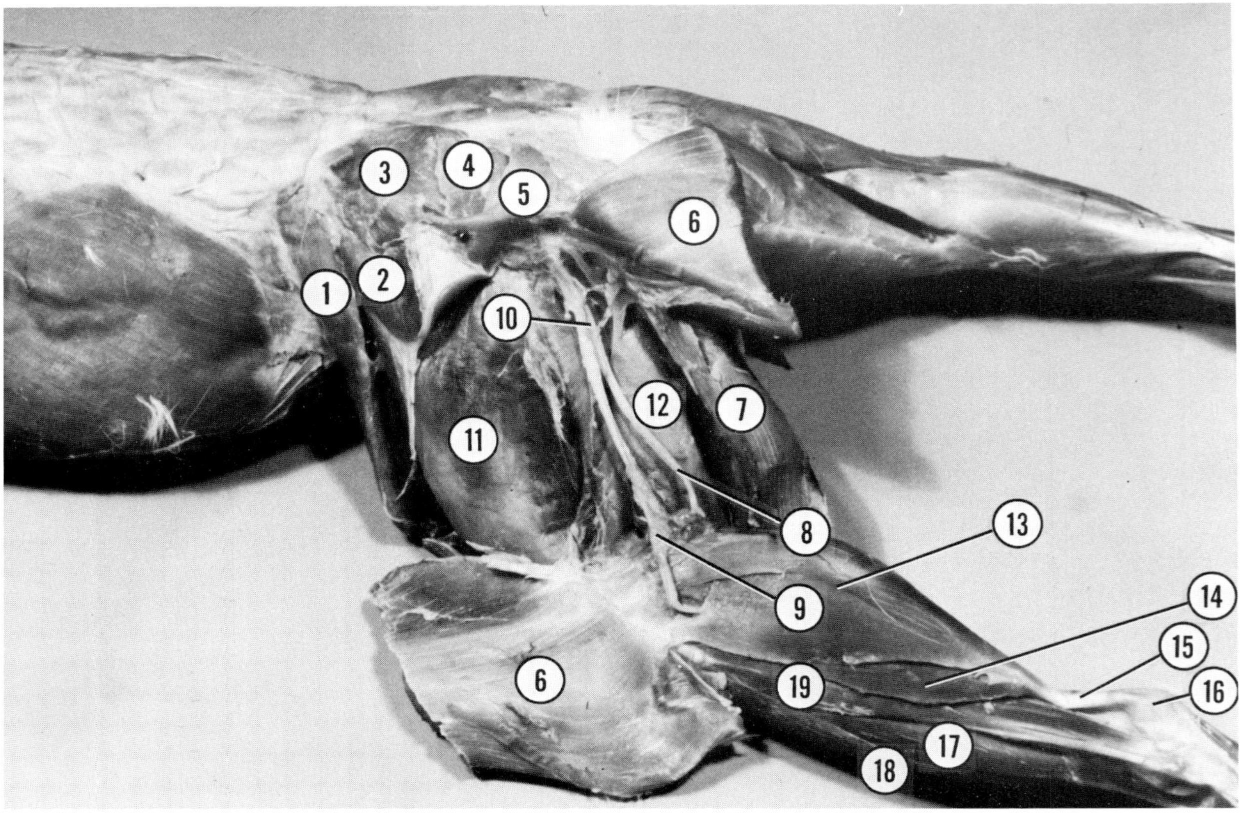

1. **Sartorius**
2. **Tensor fasciae latae**
3. **Gluteus medius**
4. **Gluteus maximus**
5. **Caudofemoralis**
6. **Biceps femoris,** transected and reflected
7. **Semitendinosus**
8. **Tibial** (TIB-ē-ul) **nerve**
9. **Common peroneal** (per´ ō-NĒ-ul) **nerve**
10. **Sciatic** (sī-AT-ik) **nerve**
11. **Vastus lateralis**
12. **Adductor femoris**
13. **Gastrocnemius** (gas´ trok-NĒ-mē´ us)
14. **Soleus** (SŌ-lē-us)
    - **A:** Extension of foot
    - **O:** Fibula, lateral surface of head
    - **I:** Calcaneus, proximal end
15. **Tendon of Achilles** (a-KIL-ēz)
16. **Calcaneus** (kal-KĀ-nē-us)
17. **Extensor digitorum longus**
    - **A:** Extends digits
    - **O:** Femur, lateral epicondyle
    - **I:** Distal phalanges, base
18. **Tibialis** (tib´ ē-Ā-lis) **anterior**
    - **A:** Flexes and laterally rotates foot
    - **O:** Tibia and fibula, proximal ends
    - **I:** First metatarsal
19. **Peroneus** (per´ ō-NĒ-us) **longus**
    - **A:** Flexes and rotates foot
    - **O:** Fibula, lateral surface of head & shaft
    - **I:** Metatarsals, base

FIGURE 25 MUSCLES OF THE LEFT SHANK - LATERAL VIEW

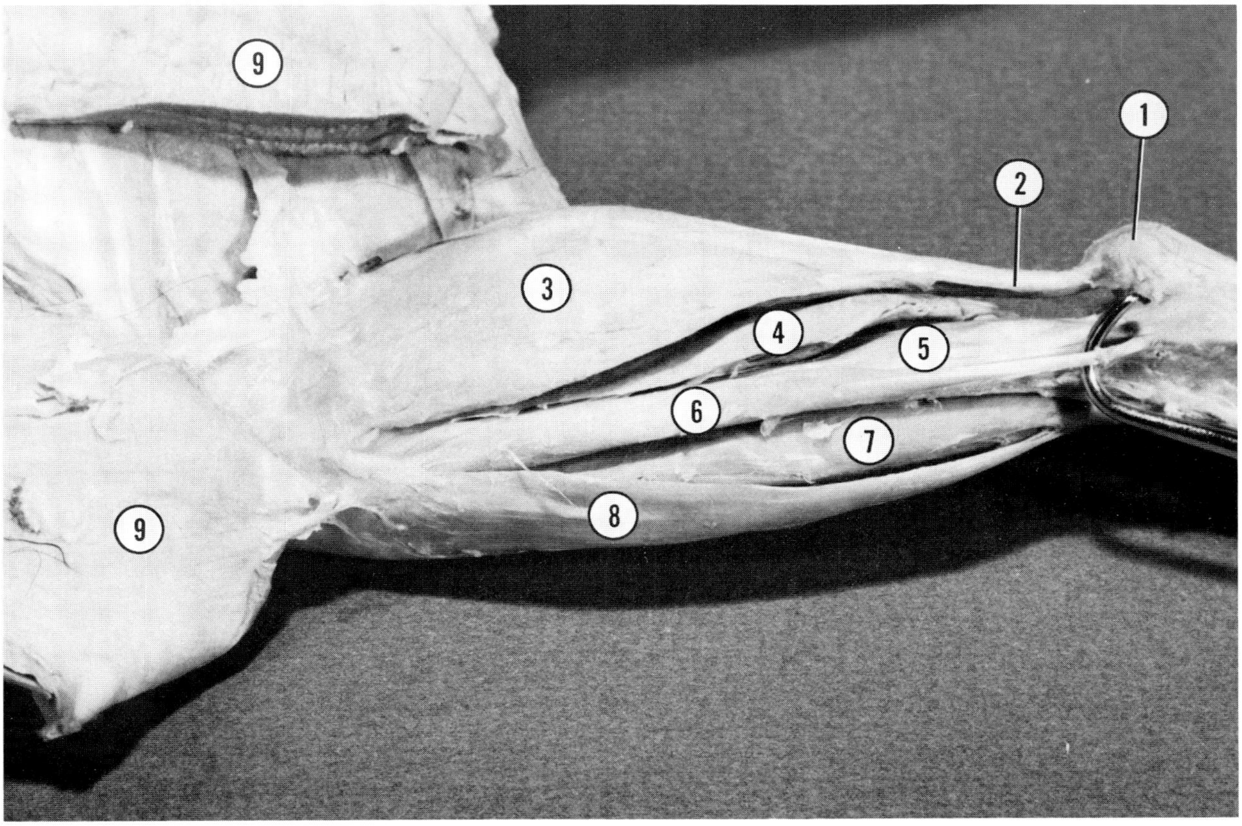

1. **Calcaneus**
2. **Tendon of Achilles**
3. **Gastrocnemius**
4. **Soleus**
5. **Peroneus** (per´ ō-NĒ-us) **brevis**
   - **A:** Extends foot
   - **O:** Fibula, distal half
   - **I:** Fifth metatarsal, base
6. **Peroneus longus**
7. **Extensor digitorum** (di-ji-TOR-um) **longus**
8. **Tibialis** (tib-ē-Ā-lis) **anterior**
9. **Biceps femoris,** transected and reflected

FIGURE **26** MUSCLES OF THE LEFT SHANK - MEDIAL VIEW

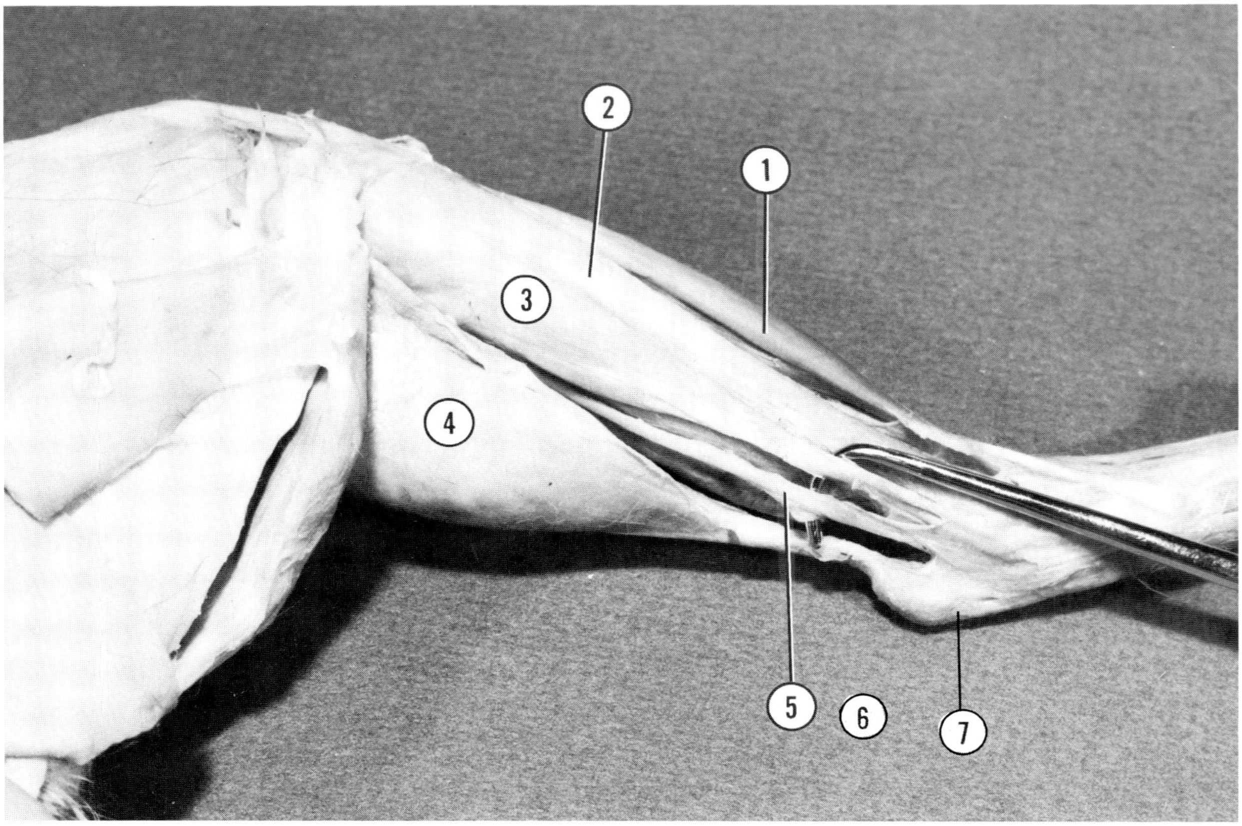

1. **Tibialis anterior**
2. **Tibia** (TIB-ē-a)
3. **Flexor digitorum longus**
   - **A:** Flexes digits and extends foot
   - **O:** Tibia, posterior surface; fibula, medial surface of head
   - **I:** Distal phalanges, base
4. **Gastrocnemius**
5. **Tibial nerve**
6. **Tendon of Achilles**
7. **Calcaneus**

FIGURE 27

ISOLATED COMPLETE CAT NERVOUS SYSTEM SHOWING ALL MAJOR NERVES - DORSAL VIEW

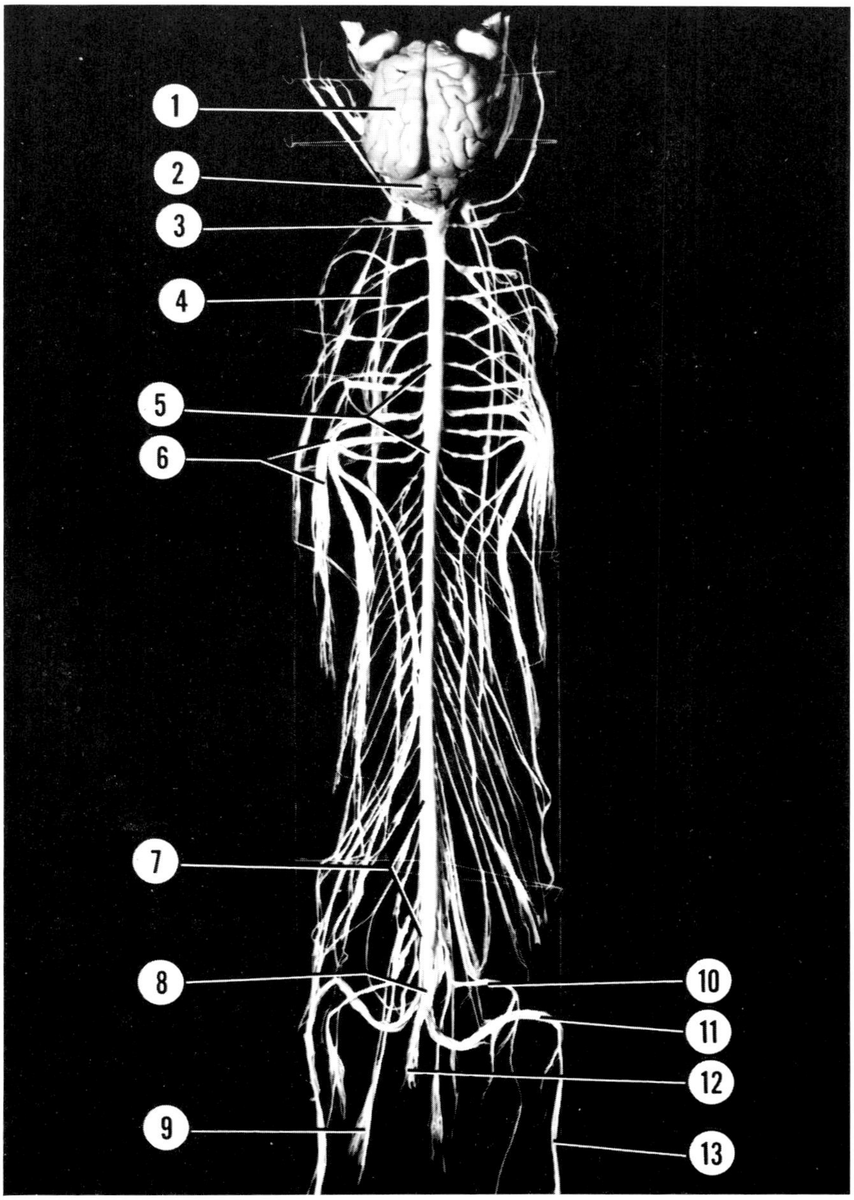

1. **Left cerebral** (SER-ē-bral) **hemisphere**
2. **Cerebellum** (ser-e-BEL-um)
3. **Medulla oblongata** (me-DULL-la ob´ long-GA-ta)
4. **Vagus** (VĀ-gus) **nerve**
5. **Cervical** (SER-vi-kul) **enlargement**
6. **Brachial** (BRĀ-kē-al) **plexus**
7. **Lumbosacral** (lum-bō-SĀ-krul) **enlargement**
8. **Filum terminale** (FĪ-lum ter-mi-NAL-ē)
9. **Inferior gluteal** (gloo-TĒ-ul) **nerve**
10. **Superior gluteal nerve**
11. **Sacral** (SĀ-krul) **plexus**
12. **Cauda equina** (KAW-da ē-KWĪ-na)
13. **Sciatic** (sī-AT-ik) **nerve**

FIGURE **28** CAT BRAIN, SPINAL CORD, AND CERVICAL NERVES - DORSAL VIEW

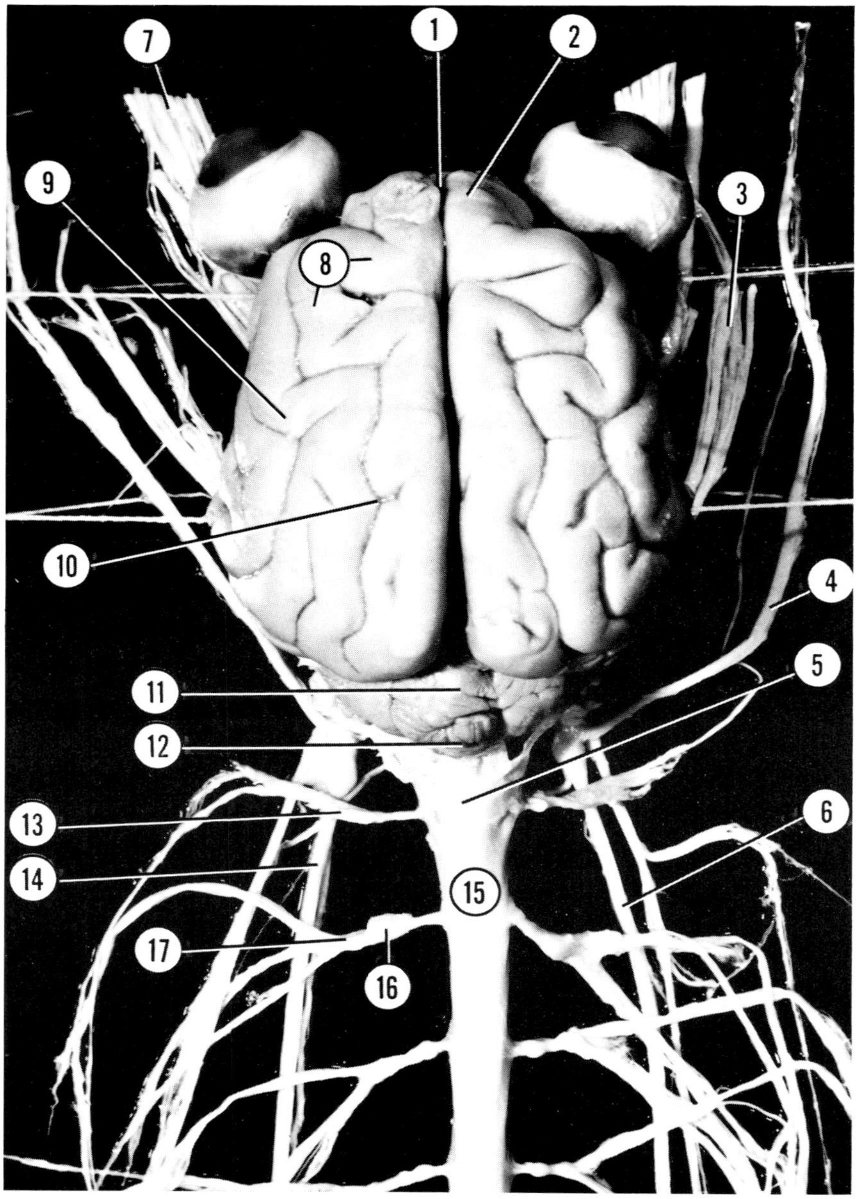

1. **Longitudinal fissure** (FISH-ur)
2. **Frontal lobe**
3. **Facial nerve**
4. **Glossopharyngeal** (glos´ ō-fa-RIN-jē-ul) **nerve**
5. **Medulla oblongata**
6. **Vagus nerve, right branch**
7. **Trigeminal** (trī-JEM-i-nal) **nerve**
8. **Gyrus** (JĪ-rus)
9. **Left cerebral hemisphere**
10. **Lateral sulcus** (SUL-kus)
11. **Cerebellum**
12. **Vermis** (VER-mis) **of cerebellum**
13. **Hypoglossal** (hī-pō-GLOS-ul) **nerve**
14. **Vagus nerve, left branch**
15. **Spinal cord**
16. **Dorsal spinal root ganglion** (GANG-lē-on)
17. **Cervical spinal nerve**

FIGURE 29 SHEEP BRAIN - DORSAL VIEW

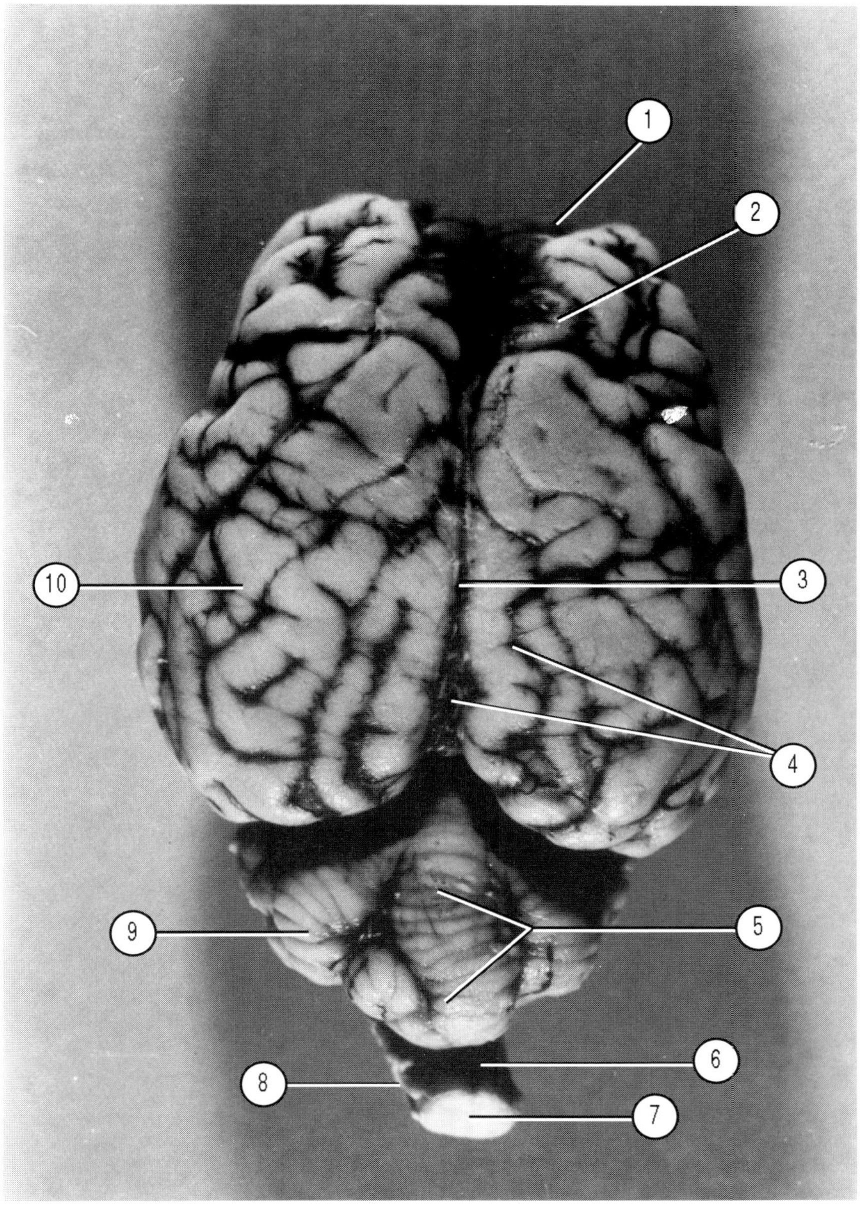

1. **Olfactory bulb**
2. **Precoronal gyrus** (JĪ-rus) **of cerebrum** (ser-Ē-brum)
3. **Median longitudinal fissure**
4. **Cerebral veins and arteries below arachnoid** (a-RAK-noyd)
5. **Cerebellar** (ser-e-BEL-ar) **vermis**
6. **Spinal cord**
7. **Central canal of spinal cord,** cut edge
8. **Spinal accessory nerve**
9. **Left cerebellar hemisphere**
10. **Left cerebral hemisphere**

FIGURE **30** SHEEP BRAIN - LATERAL VIEW

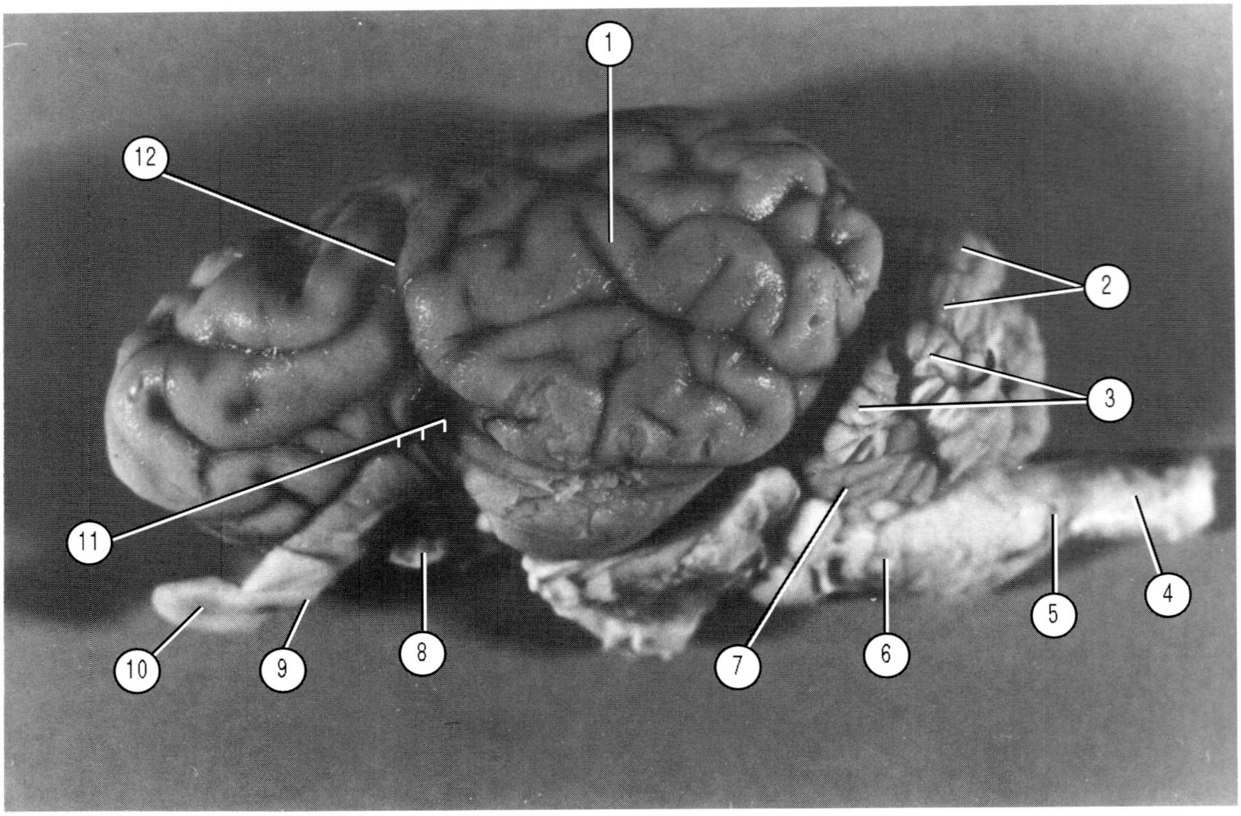

1. **Left cerebral hemisphere**
2. **Cerebellar vermis**
3. **Paraflocculus** (par-a-FLOK-ū-lus)
4. **Spinal cord**
5. **Medulla oblongata** (me-DULL-la ob´ long-GA-ta)
6. **Pons**
7. **Flocculus** (FLOK-ū-lus)
8. **Optic nerve**
9. **Olfactory tract**
10. **Olfactory bulb**
11. **Branches of cerebral artery**
12. **Sylvian gyrus** (sil-VĒ-an JĪ-rus)

FIGURE 31 SHEEP BRAIN - VENTRAL VIEW

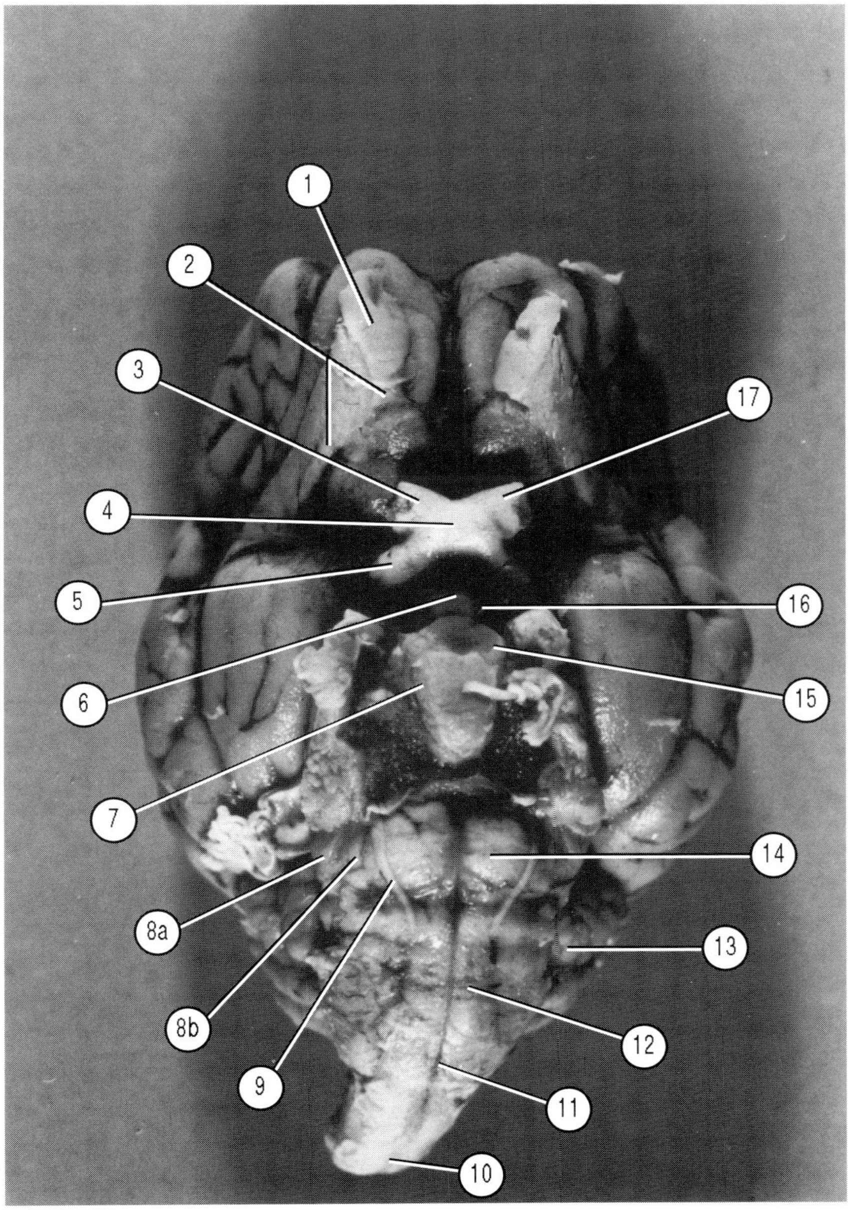

1. **Olfactory bulb**
2. **Olfactory tracts** (**Medial** and **Lateral**)
3. **Optic nerve, right**
4. **Optic chiasma** (kī-AS-ma)
5. **Optic tract**
6. **Infundibulum** (in´ fun-DIB-yoo-lum)
7. **Midbrain**

8a. **Trigeminal** (trī-GEM-e-nul) **nerve, sensory root**

8b. **Trigeminal nerve, motor root**

9. **Abducens** (ab-DOO-cenz) **nerve**
10. **Spinal cord**
11. **Ventral median fissure**
12. **Medulla oblongata**
13. **Cerebellar flocculus**
14. **Pons**
15. **Mammillary body**
16. **Tuber cinereum** (TŪ-ber sin-er-Ē-um)
17. **Optic nerve, left**

FIGURE 32 SHEEP BRAIN VIEWED IN SAGITTAL SECTION

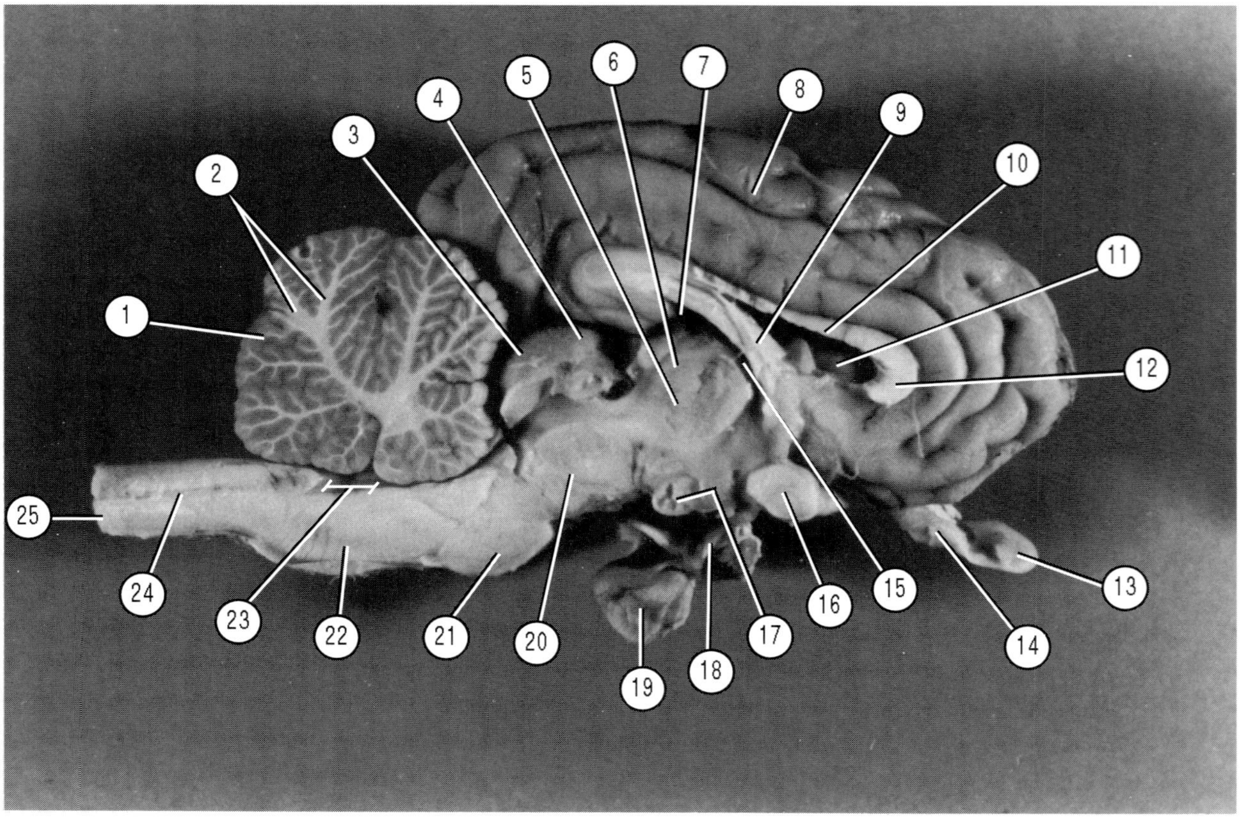

1. **Cerebellum**
2. **Arbor vitae** (AR-bor VĪ-tē)
3. **Superior colliculus** (kol-LIK-ū-lus)
4. **Pineal** (pī-NĒ-ul) **body**
5. **Massa intermedia**
6. **Thalamus** (THAL-a-mus)
7. **Third ventricle**
8. **Cerebral hemisphere, left**
9. **Fornix**
10. **Body of corpus callosum** (KOR-pus ka-LŌ-sum)
11. **Lateral ventricle, left** (septum pellucidum removed to expose ventricle)
12. **Genu** (JĒ-noo) **of corpus callosum**
13. **Olfactory bulb**
14. **Olfactory tract**
15. **Interventricular foramen**
16. **Optic chiasma**
17. **Mammillary body**
18. **Infundibulum**
19. **Hypophysis** (hī-POF-i-sis)
20. **Midbrain**
21. **Pons**
22. **Medulla oblongata**
23. **Fourth ventricle**
24. **Central canal**
25. **Spinal cord,** cut edge

FIGURE 33 SPINAL CORD: CERVICAL REGION SEEN IN PANORAMIC VIEW-TRANSVERSE SECTION

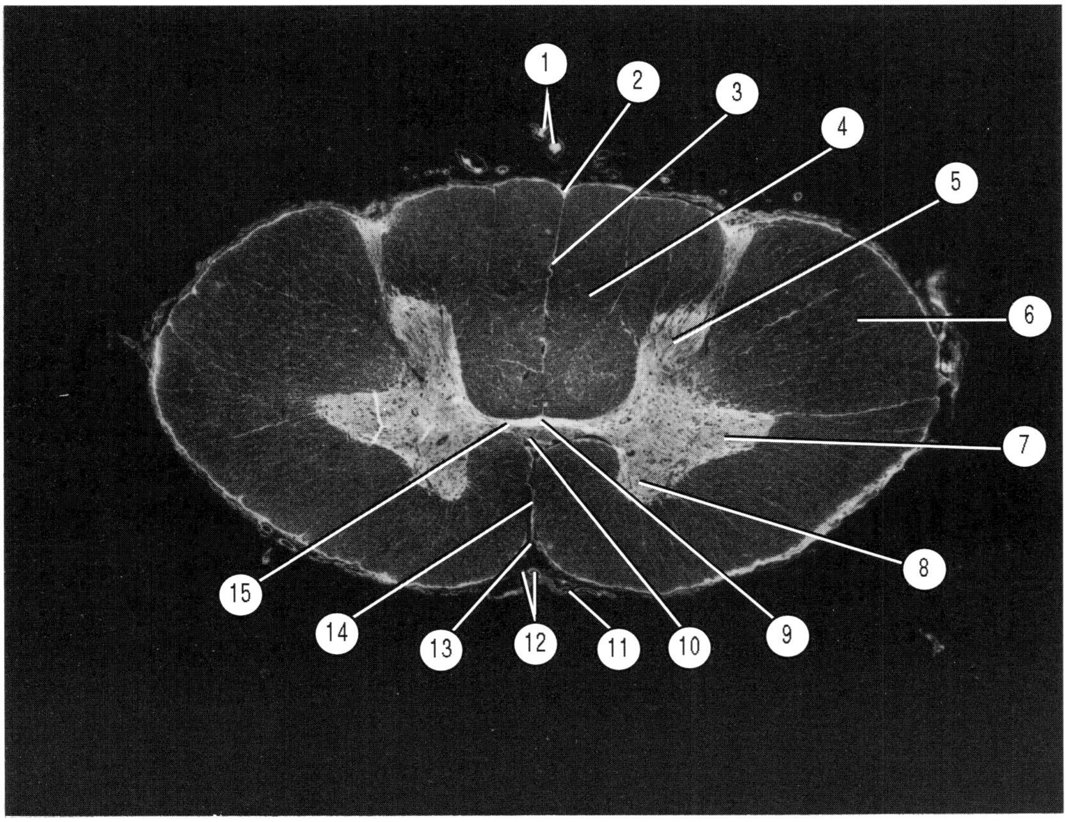

1. **Posterior spinal vessels (veins)**
2. **Posterior median sulcus**
3. **Posterior median septum**
4. **Posterior medial white column**
5. **Posterior gray horn**
6. **Lateral white column**
7. **Anterior lateral gray horn containing motor nuclei**
8. **Anterior medial gray horn**
9. **Central canal**
10. **Anterior white commissure**
11. **Arachnoid** (a-RAK-noyd)
12. **Anterior spinal artery and vein**
13. **Anterior median sulcus and pia mater**
14. **Anterior median fissure**
15. **Gray commissure**

FIGURE 34 SHEEP EYE DISSECTED TO SHOW INTERNAL STRUCTURE FRONTAL (CORONAL) SECTION

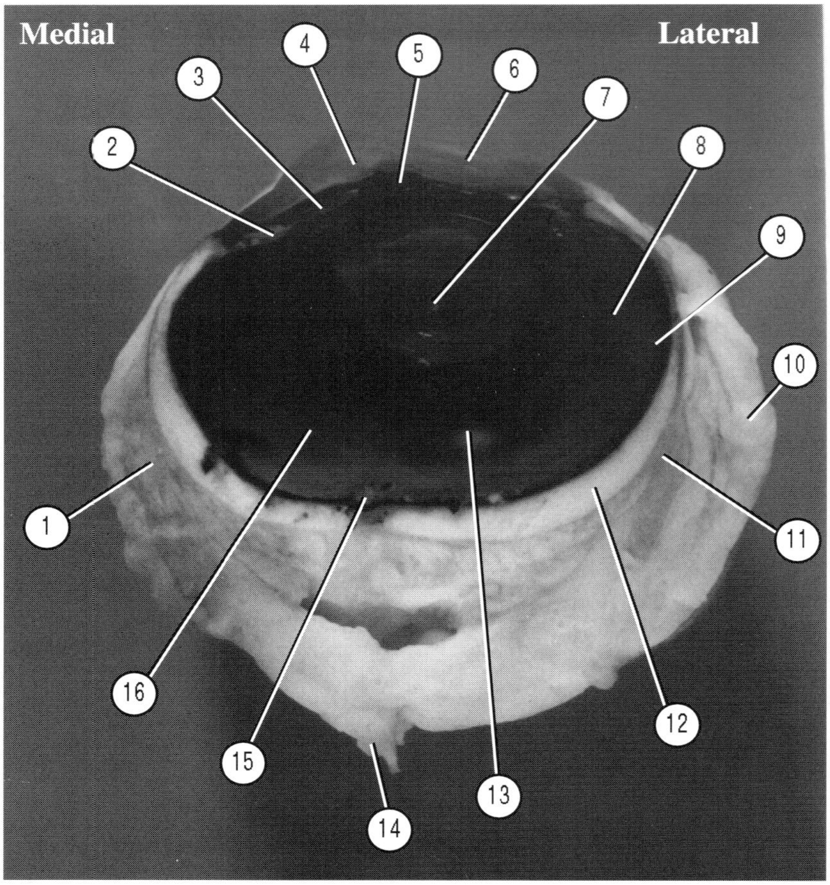

1. **Medial rectus muscle,** cut
2. **Ciliary body**
3. **Iris**
4. **Anterior chamber**
5. **Pupil**
6. **Cornea**
7. **Lens**
8. **Neural tunic (retina)**
9. **Vascular tunic (choroid)**
10. **Orbital fat**
11. **Lateral rectus muscle,** cut
12. **Fibrous tunic (sclera)**
13. **Posterior cavity (chamber of the vitreous body) containing the vitreous body**
14. **Optic nerve surrounded by orbital fat**
15. **Optic disc**
16. **Retinal blood vessels**

FIGURE 35 SHEEP EYE, POSTERIOR HALF DISSECTED - TRANSVERSE SECTION

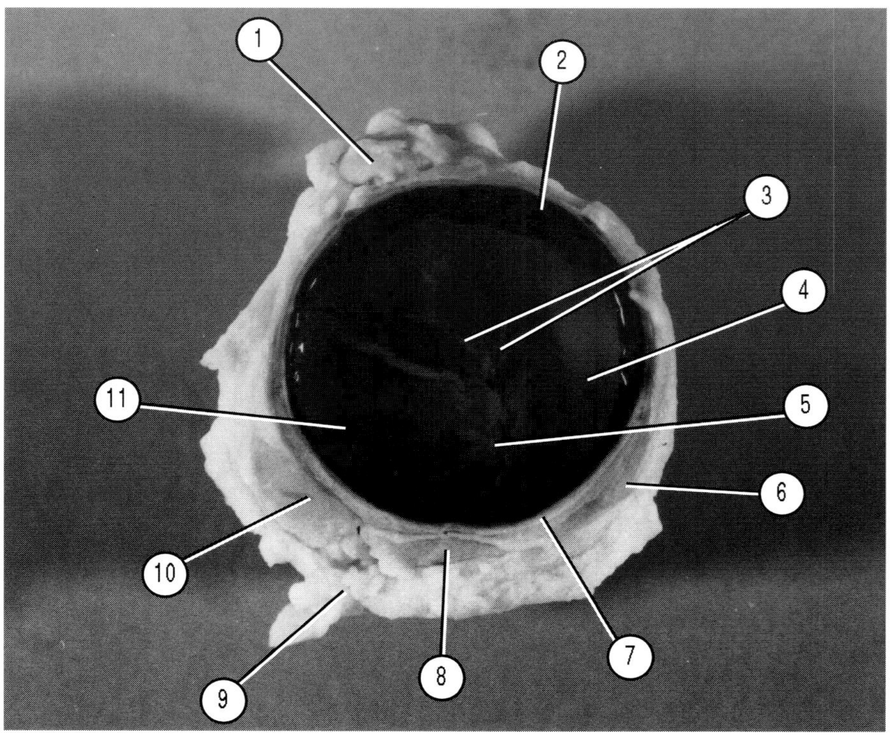

1. **Dorsal rectus muscle,** cut edge
2. **Vascular tunic (choroid)**
3. **Retinal blood vessels**
4. **Neural tunic (retina)**
5. **Optic disc**
6. **Medial rectus muscle,** cut edge
7. **Fibrous tunic (sclera)**
8. **Ventral rectus muscle,** cut edge
9. **Orbital fat**
10. **Lateral rectus muscle,** cut edge
11. **Vitreous body contained in posterior cavity**

FIGURE 36 SHEEP HEART DISSECTED TO SHOW EXTERNAL STRUCTURE - VENTRAL VIEW

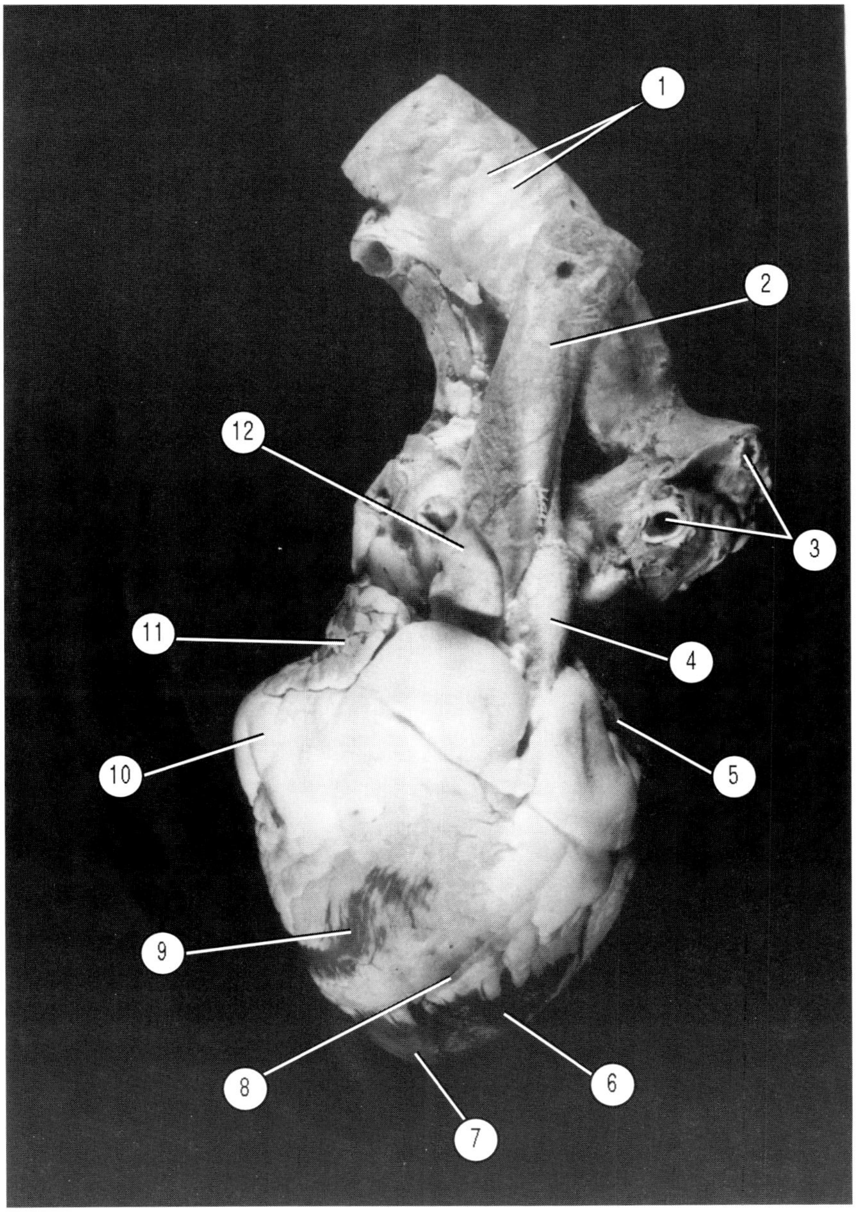

1. **Trachea**
2. **Aorta**
3. **Lobar bronchi to left lung**
4. **Pulmonary trunk artery**
5. **Auricle of left atrium**
6. **Epicardium of left ventricle**
7. **Apex**
8. **Left coronary artery and vein within ventral interventricular sulcus**
9. **Epicardium of right ventricle**
10. **Fat deposit associated with epicardium**
11. **Auricle of right atrium**
12. **Innominate (Brachiocephalic) artery**

FIGURE 37 SHEEP HEART DISSECTED TO SHOW EXTERNAL STRUCTURE - DORSAL VIEW

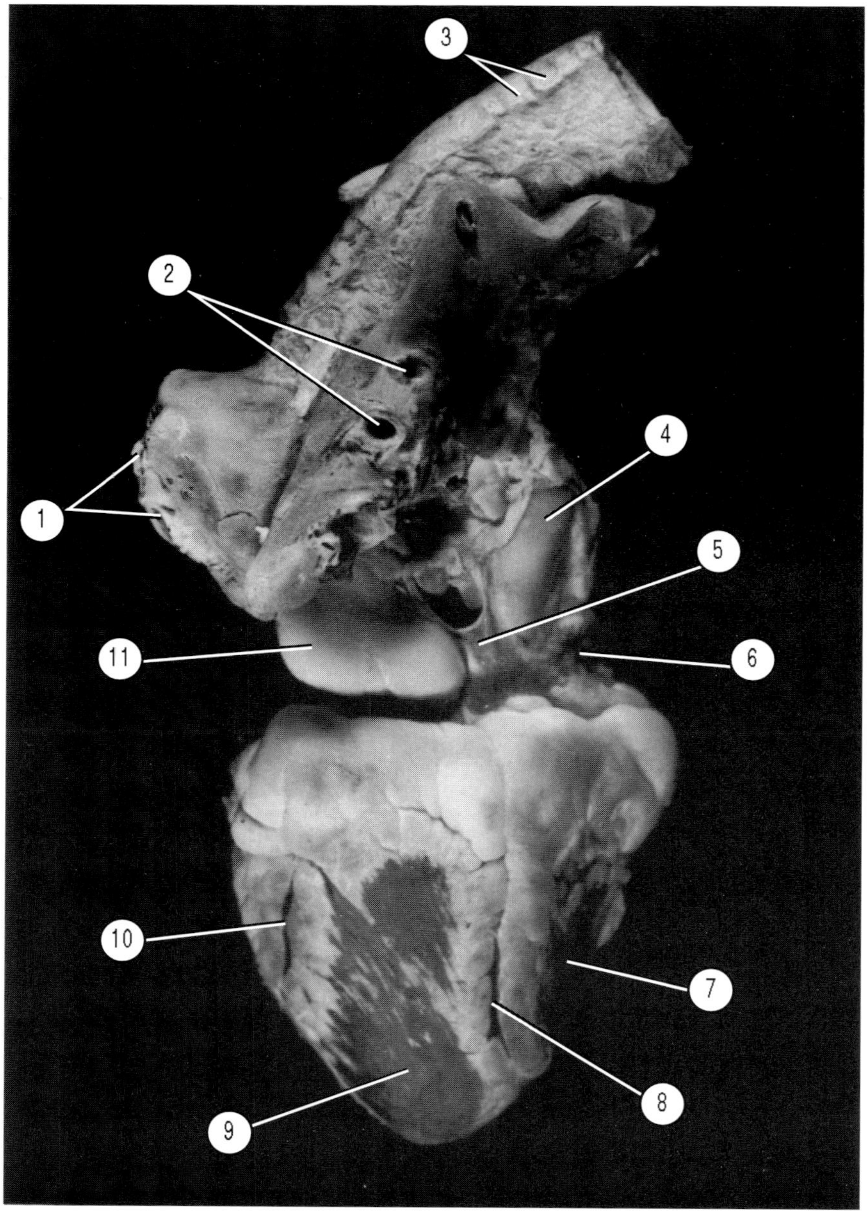

1. **Pulmonary arteries**
2. **Pulmonary veins**
3. **Cartilage of trachea**
4. **Innominate (Brachiocephalic) artery**
5. **Posterior vena cava (Postcava)**
6. **Auricle of right atrium**
7. **Myocardium of right ventricle**
8. **Right coronary artery and vein**
9. **Myocardium of left ventricle**
10. **Left coronary artery and vein**
11. **Auricle of left atrium**

FIGURE **38** SHEEP HEART DISSECTED TO SHOW INTERNAL STRUCTURE - VENTRAL VIEW

1. **Anterior vena cava**
2. **Innominate (Brachiocephalic) artery**
3. **Arch of aorta**
4. **Ductus arteriosus**
5. **Pulmonary trunk**
6. **Left atrium**
7. **Pectinate muscles**
8. **Cusp of bicuspid (Mitral) valve**
9. **Chordae tendineae of left ventricle**
10. **Myocardium of left ventricle**
11. **Papillary muscle**
12. **Apex**
13. **Fat associated with epicardium**
14. **Interventricular septum**
15. **Myocardium of right ventricle**
16. **Muscular moderator band**
17. **Chordae tendineae of right ventricle**
18. **Tricuspid valve**
19. **Pulmonary semilunar valve**
20. **Auricle of right atrium**

FIGURE 39 DETAIL OF HEART IN SITU WITH PERICARDIUM REFLECTED - VENTRAL ASPECT

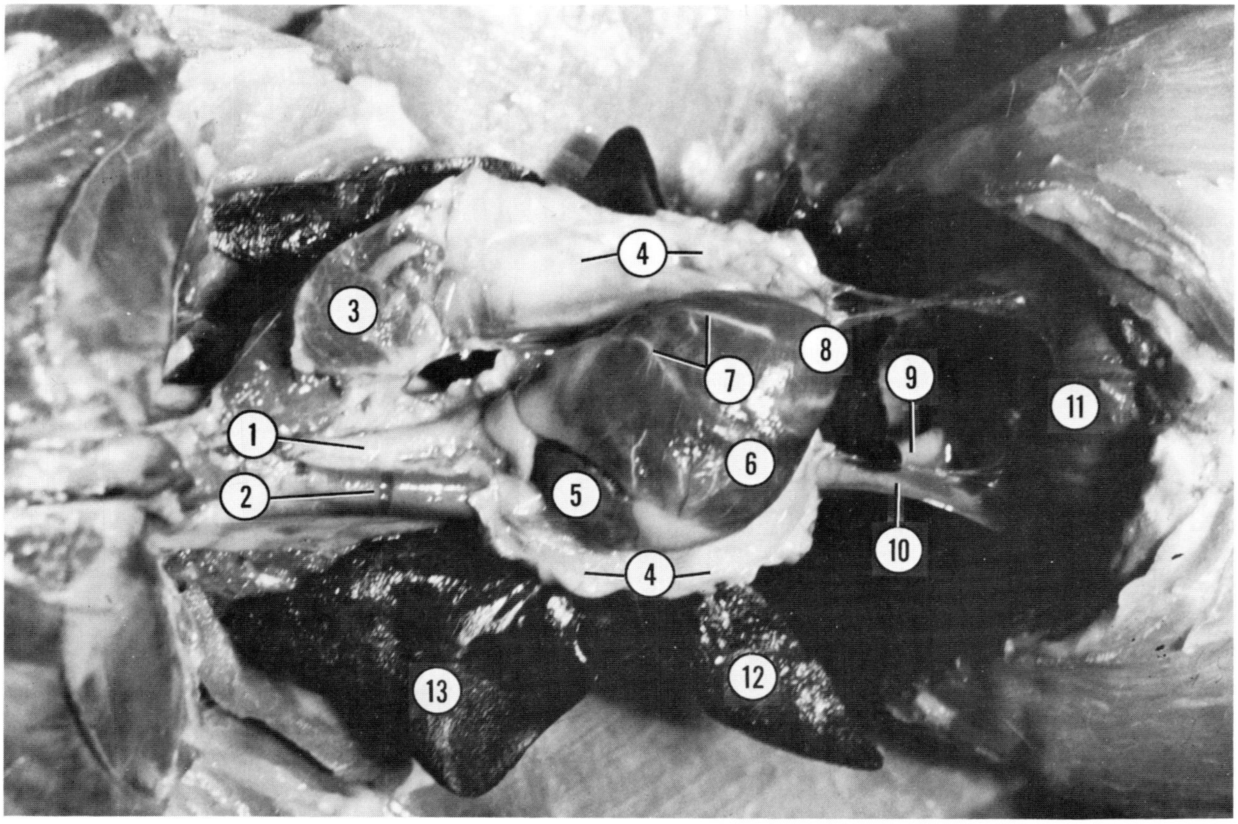

1. **Innominate** (i-NOM-i-nut) or **Brachiocephalic** (brā-kē-ō-se-FAL-ik) **artery**
2. **Precava** (prē-KĀ-va) **vein (Anterior vena cava)** (VĒ-na KĀ-va)
3. **Thymus** (THĪ-mus) **gland**
4. **Pericardium,** cut and reflected
5. **Auricle of right atrium**
6. **Right ventricle**
7. **Coronary artery and vein**
8. **Apex of heart**
9. **Thoracic** (thō-RAS-ik) **aorta**
10. **Postcava** (post-KĀ-va) or **Posterior vena cava**
11. **Diaphragm** (DĪ-a-fram)
12. **Right middle lung lobe**
13. **Right anterior lung lobe**

FIGURE 40 VIEW OF THORACIC VISCERA AFTER REMOVAL OF PERICARDIUM, THYMUS, AND FASCIA

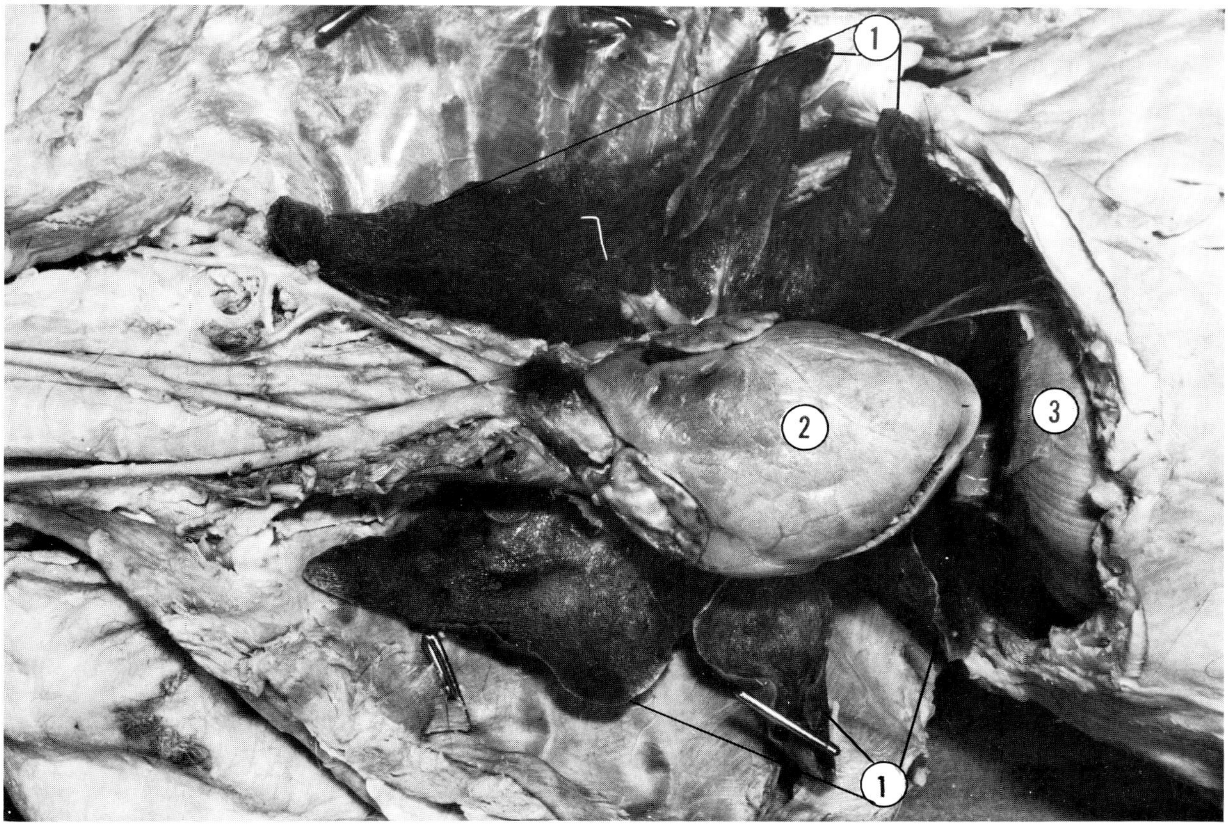

1. **Lungs**
2. **Heart**
3. **Diaphragm** (DI-a-fram)

FIGURE **41** CAT HEART - SURFACE FEATURES

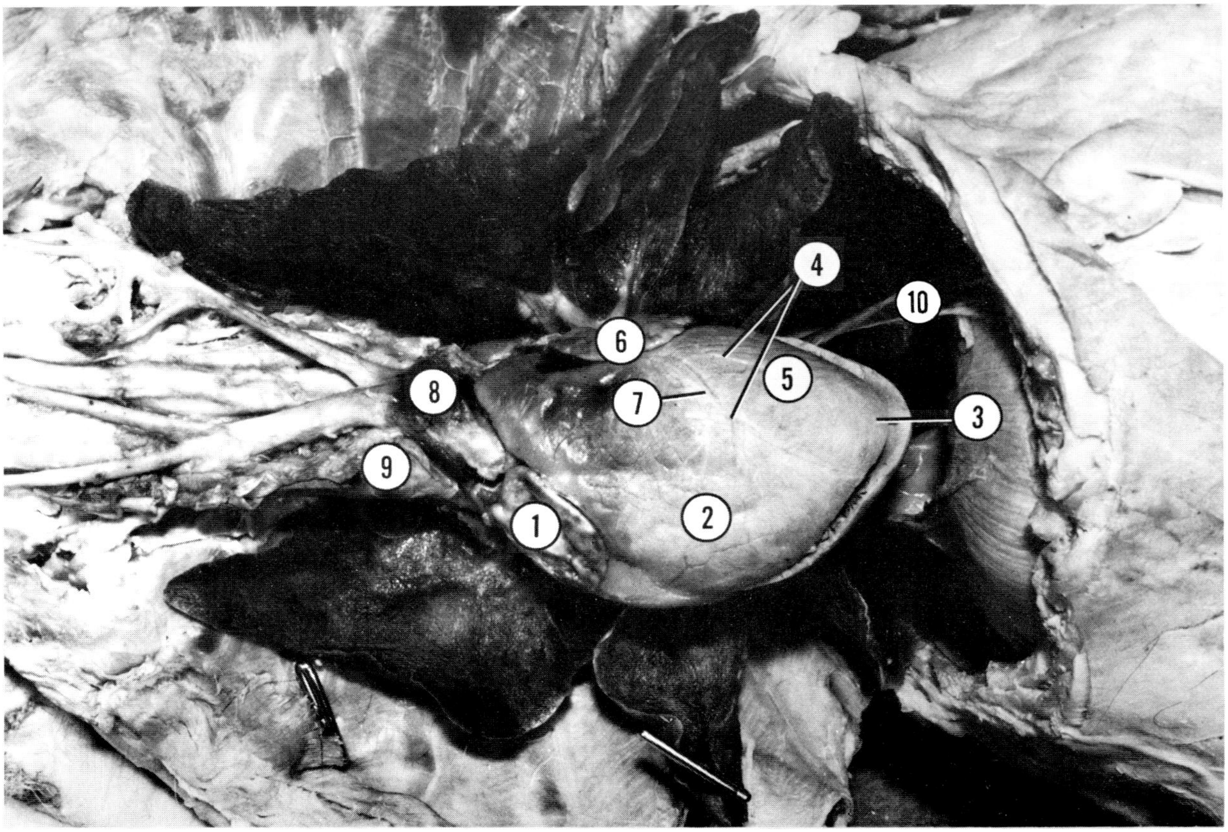

1. **Auricle of right atrium**
2. **Right ventricle**
3. **Apex of heart**
4. **Left coronary** (KOR-ō-ner-ē) **artery and vein**
5. **Left ventricle**
6. **Auricle of left atrium**
7. **Interventricular** (in´ tur-ven-TRIK-yoo-lur) **groove**
8. **Aortic** (ā-OR-tik) **arch**
9. **Precava vein (Anterior vena cava)**
10. **Pulmonary** (PUL-mō-ner´ ē) **ligament**

FIGURE 42 BLOOD VESSELS OF THE NECK REGION AND THORACIC CAVITY (LUNGS AND THYMUS GLAND REMOVED)

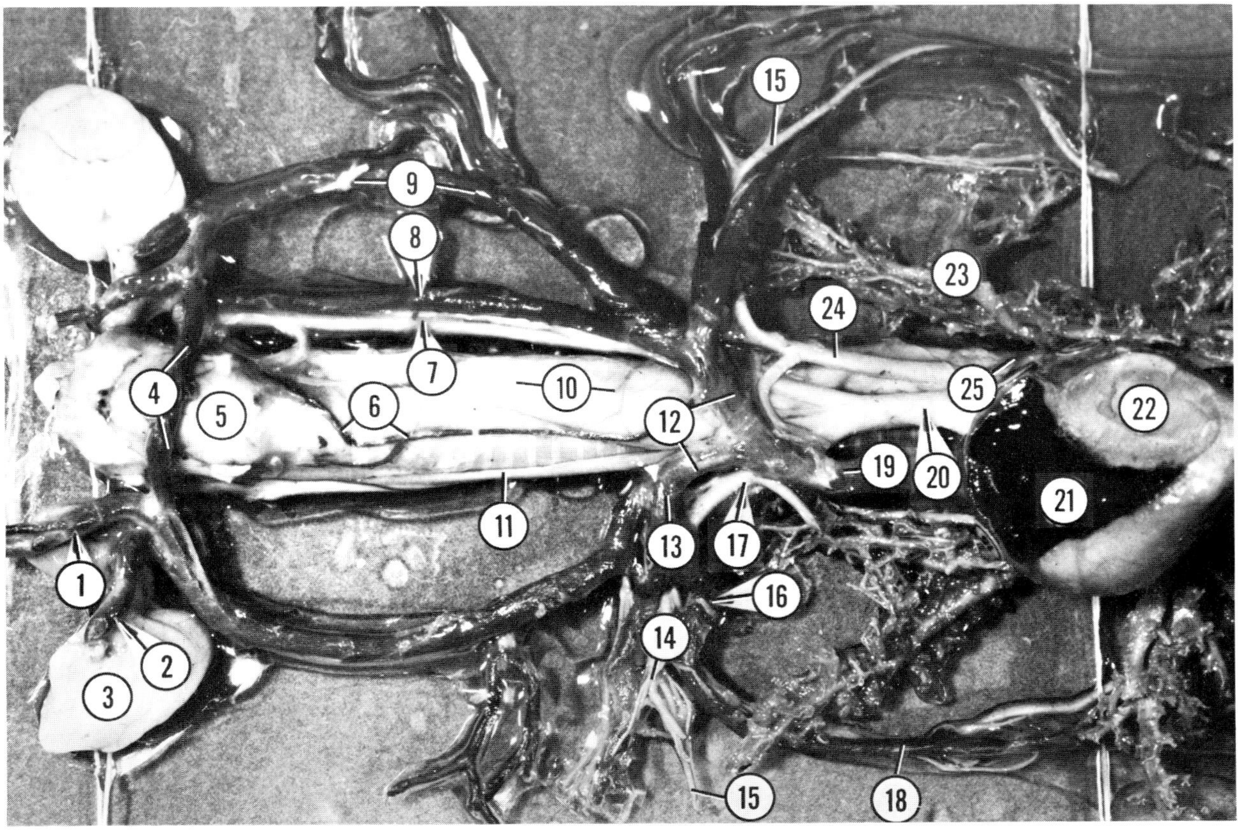

1. **Anterior facial vein**
2. **Posterior facial vein**
3. **Submandibular salivary gland** (sub´ man-DIB-yoo-lur SAL-i-ver´ ē)
4. **Transverse jugular** (JUG-yoo-lar) **vein**
5. **Thyroid** (THĪ-royd) **cartilage**
6. **Inferior thyroid vein**
7. **Vagus** (VĀ-gus) **nerve, left branch** connected to left **common carotid** artery
8. **Internal jugular vein**
9. **External jugular vein**
10. **Esophagus** (ē-SOF-a-gus)
11. **Right common carotid artery**
12. **Innominate (Brachiocephalic) vein**
13. **Right subclavian vein**
14. **Right axillary** (AK-si-ler´ ē) **artery**
15. **Left brachial** (BRĀ-kē-al) **artery**
16. **Right axillary vein**
17. **Right subclavian artery**
18. **Right brachial vein**
19. **Precava (Anterior vena cava)**
20. **Innominate (Brachiocephalic) artery**
21. **Interior of right ventricle**
22. **Interior of left ventricle**
23. **Pulmonary arteries and veins**
24. **Left subclavian artery**
25. **Aortic arch**

FIGURE 43 ARTERIES AND VEINS OF THE NECK - VENTRAL VIEW

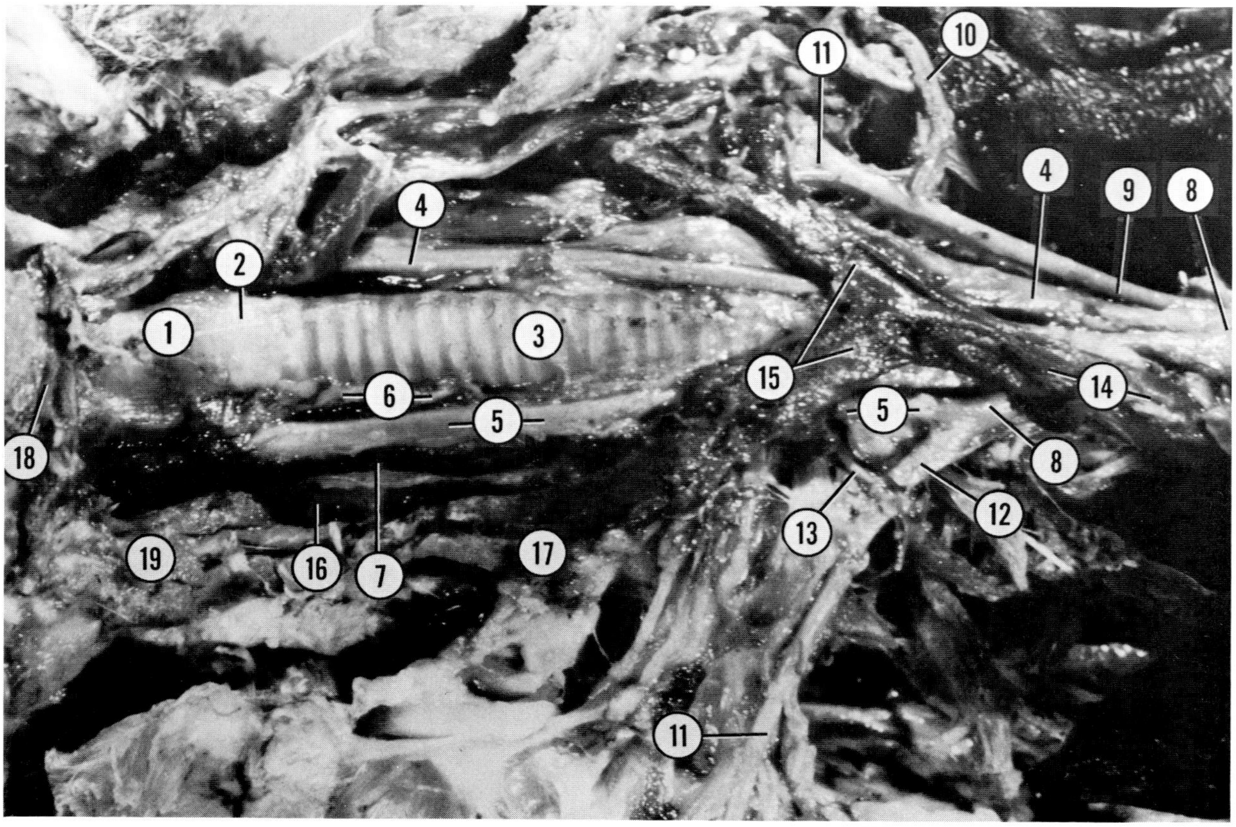

1. **Thyroid cartilage of larynx** (LAR-inks)
2. **Cricoid** (KRĪ-koyd) **cartilage of larynx**
3. **Trachea**
4. **Left common carotid artery**
5. **Right common carotid artery**
6. **Vagus nerve, right branch**
7. **Right tracheal lymphatic** (lim-FAT-ik) **duct**
8. **Innominate (Brachiocephalic) artery**
9. **Left subclavian artery**
10. **Internal mammary** (mam-ar-ē) **artery**
11. **Left axillary artery**
12. **Right subclavian artery**
13. **Thyrocervical** (thī´ rō-SER-vi-kul) **artery**
14. **Precava vein (Anterior vena cava)**
15. **Right and left innominate veins**
16. **Right internal jugular vein**
17. **Right external jugular vein**
18. **Transverse jugular vein**
19. **Medial retropharyngeal** (re-trō-fa-RIN-jē-ul) **lymph node**

FIGURE 44 BLOOD VESSELS OF THE ABDOMINAL VISCERA - VENTRAL VIEW (GREATER OMENTUM, SMALL AND LARGE INTESTINES, AND LIVER REMOVED)

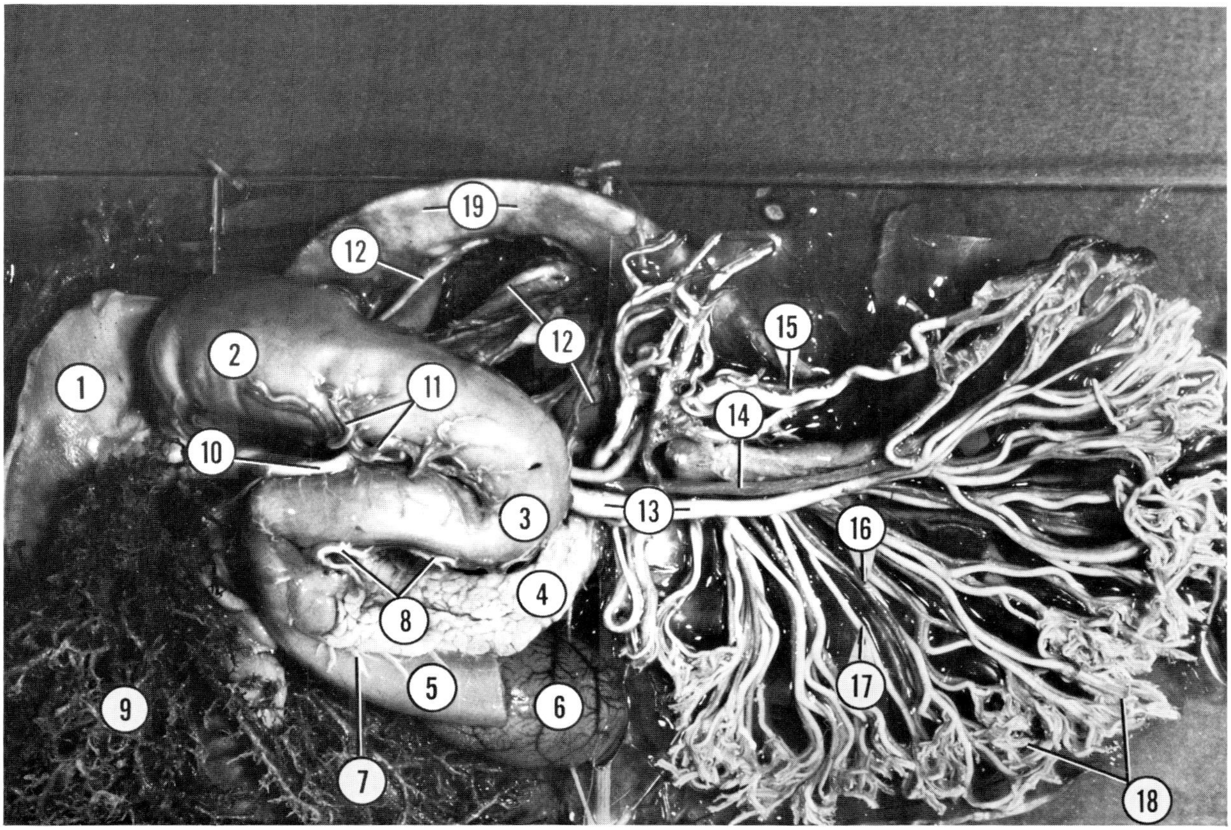

1. **Diaphragm**
2. **Fundic** (FUN-dik) **portion of stomach**
3. **Pyloric** (pī-LOR-ik) **region of stomach**
4. **Pancreas** (PAN-krē-as)
5. **Duodenum** (doo´ ō-DĒ-num)
6. **Right kidney**
7. **Duodenal** (doo´ ō-DĒ-nal) **branches of pancreaticoduodenal** (pan´ krē-AT-i-kō-doo´ ō-DĒ-nul) **arteries and veins**
8. **Pancreatic** (pan´ krē-AT-ik) **branches of pancreaticoduodenal arteries and veins**
9. **Hepatic** (he-PAT-ik) **arteries and veins of liver**
10. **Branch of the celiac** (SĒ-lē-ak) **trunk artery**
11. **Left gastric** (GAS-trik) **artery**
12. **Splenic** (SPLEN-ik) **arteries and veins**
13. **Superior mesenteric** (mez´ un-TER-ik) **artery**
14. **Superior mesenteric vein**
15. **Middle colic** (KOL-ik) **artery**
16. **Intestinal branches of superior mesenteric vein**
17. **Intestinal branches of superior mesenteric artery**
18. **Ileal** (IL-ē-ul) **branches with vasa recta** (VĀ-sa REK-ta)
19. **Spleen**

FIGURE **45** ABDOMINAL BLOOD VESSELS POSTERIOR TO THE KIDNEY

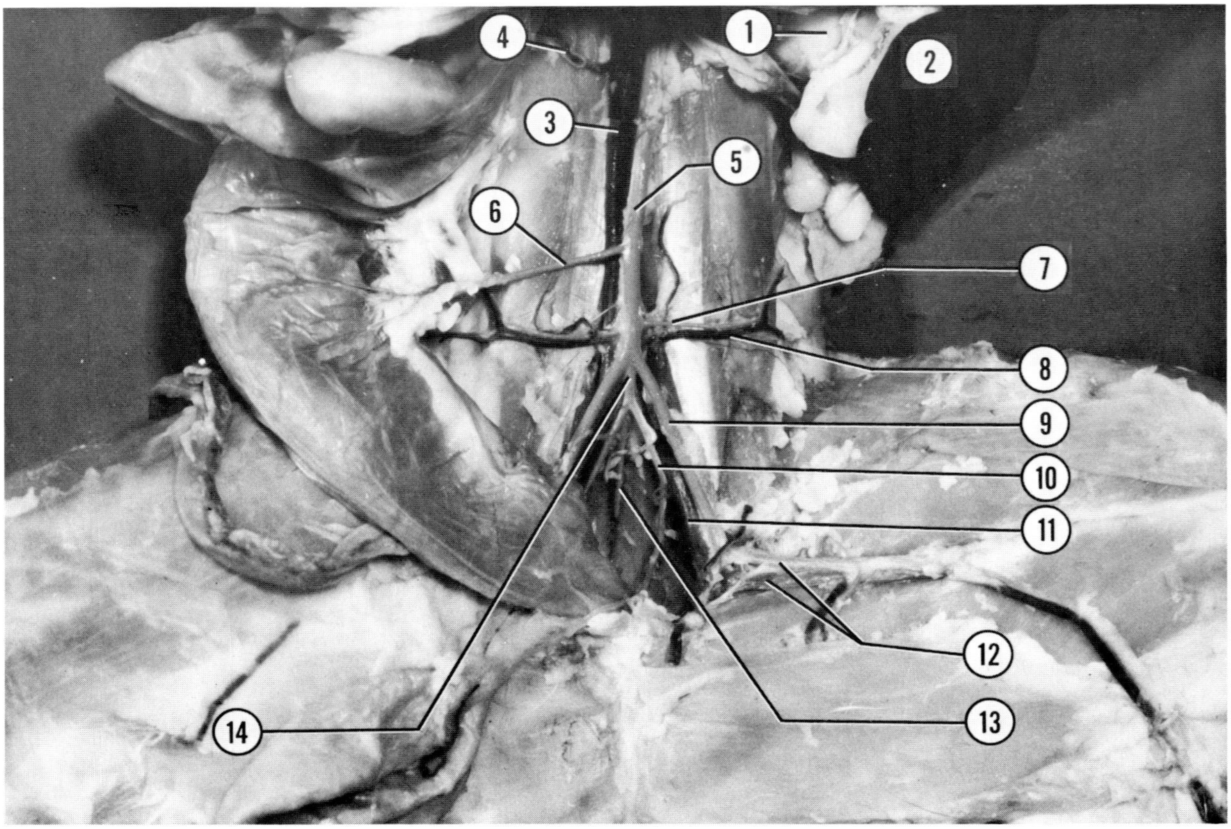

1. **Left kidney, posterior border**
2. **Spleen**
3. **Postcava (Posterior vena cava)**
4. **Internal spermatic** (sper-MAT-ik) **vein or Ovarian** (ō-VAR-ē-an) **vein**
5. **Descending abdominal** (ab-DOM-i-nal) **aorta**
6. **Posterior mesenteric artery and vein**
7. **Iliolumbar** (il´ ē-ō-LUM-bar) **or Deep circumflex** (SUR-kum-flex) **artery**
8. **Iliolumbar (Deep circumflex) vein**
9. **Left external iliac** (IL-ē-ak) **artery**
10. **Left internal iliac artery**
11. **Common iliac vein**
12. **Femoral** (FEM-or-ul) **artery and vein**
13. **Middle sacral** (SĀ-krul) **artery and vein**
14. **Common iliac artery**

FIGURE **46** BLOOD VESSELS OF THE GROIN AND THIGH - LEFT HINDLIMB

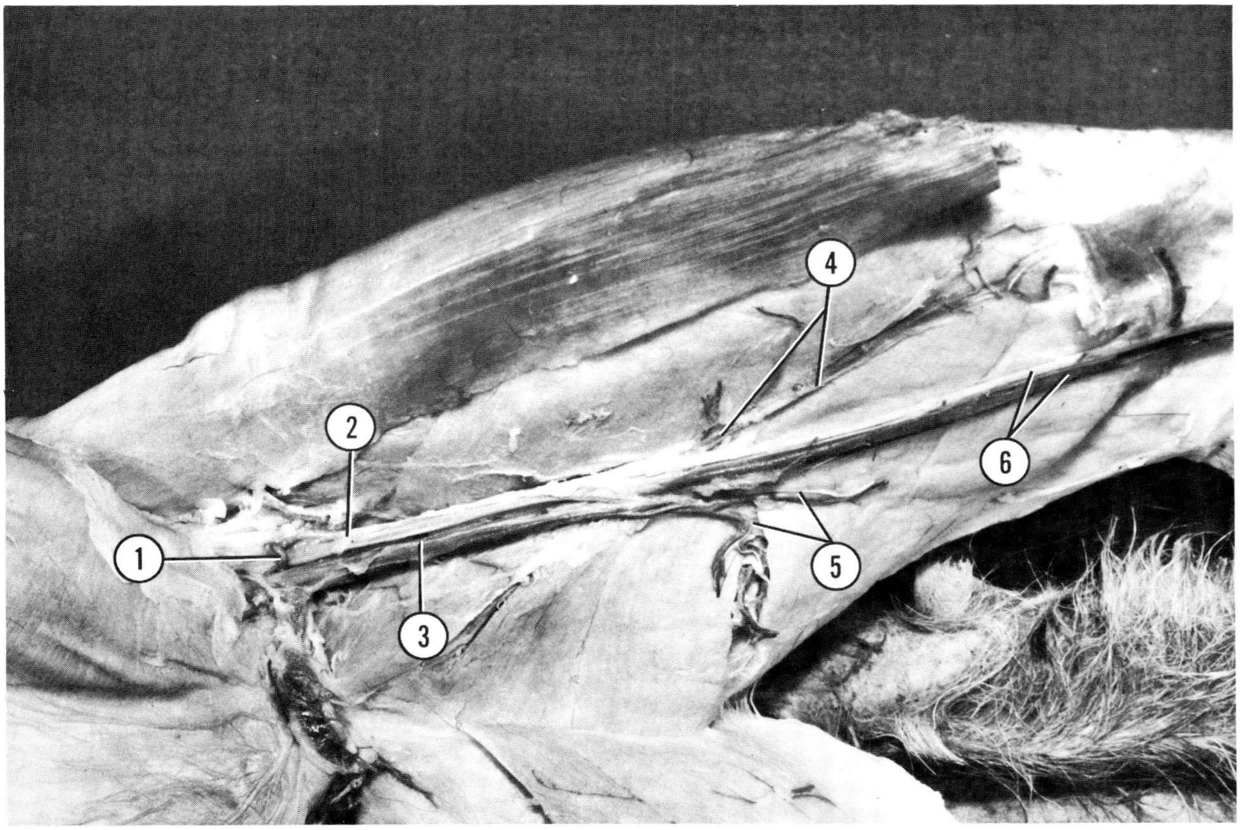

1. **Left lateral femoral circumflex artery and vein**
2. **Left femoral artery and femoral nerve**
3. **Left femoral vein**
4. **Descending genicular** (je-NIK-yoo-lur) **artery and vein**
5. **Muscular branches of left femoral artery and vein**
6. **Left saphenous** (SAF-e-nus) **artery and vein**

FIGURE 47 THE RESPIRATORY SYSTEM

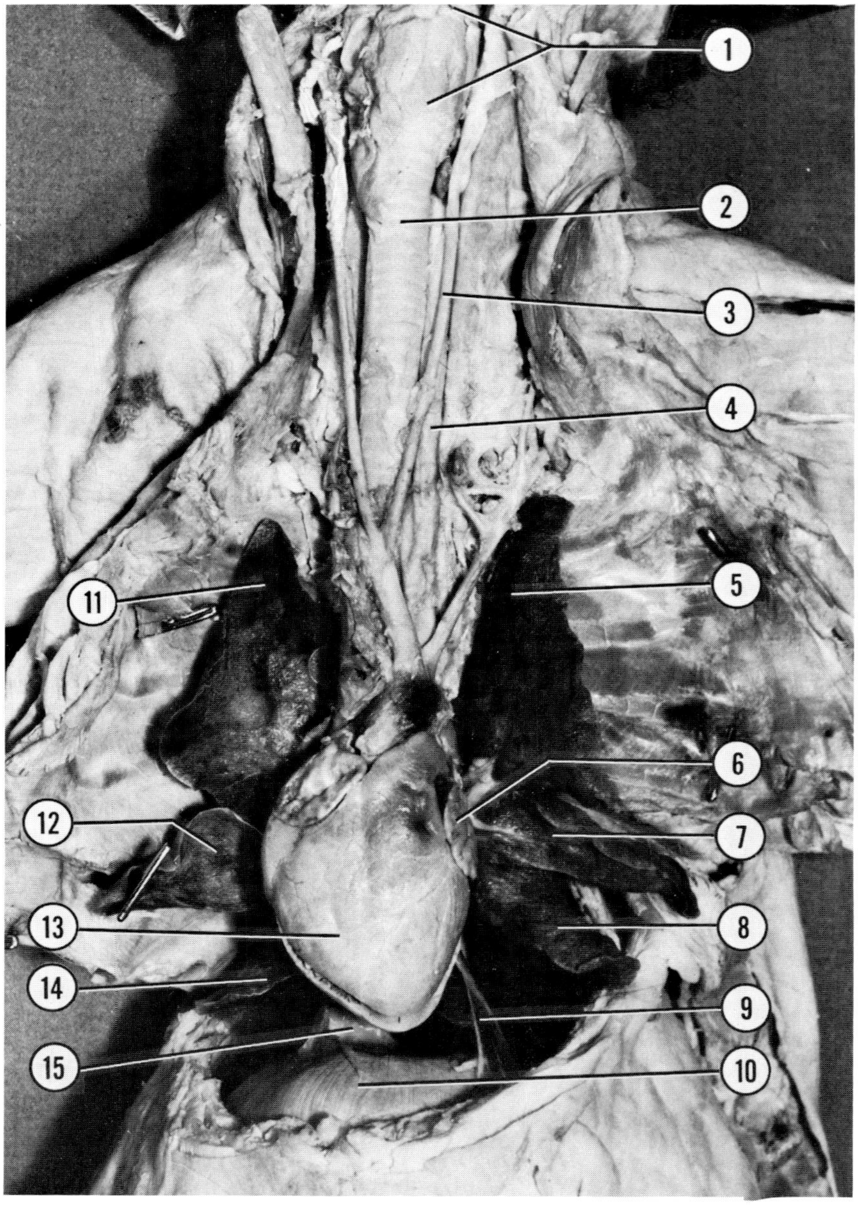

1. **Larynx** (LAR-inks)
2. **Trachea** (TRĀ-kē-a)
3. **Left common carotid artery**
4. **Esophagus** (ē-SOF-a-gus)
5. **Left anterior lung lobe**
6. **Auricle of left atrium**
7. **Left middle lung lobe**
8. **Left posterior (Diaphragmatic)** (dī-a-frag-MAT-ik) **lung lobe**
9. **Pulmonary** (PUL-mō-ner´ ē) **ligament**
10. **Diaphragm** (DĪ-a-fram)
11. **Right anterior lung lobe**
12. **Right middle lung lobe**
13. **Right ventricle**
14. **Right posterior (Diaphragmatic) lung lobe**
15. **Caval** (KĀ-val) **fold**

FIGURE **48** THORACIC VISCERA REMOVED TO SHOW LUNG DETAIL

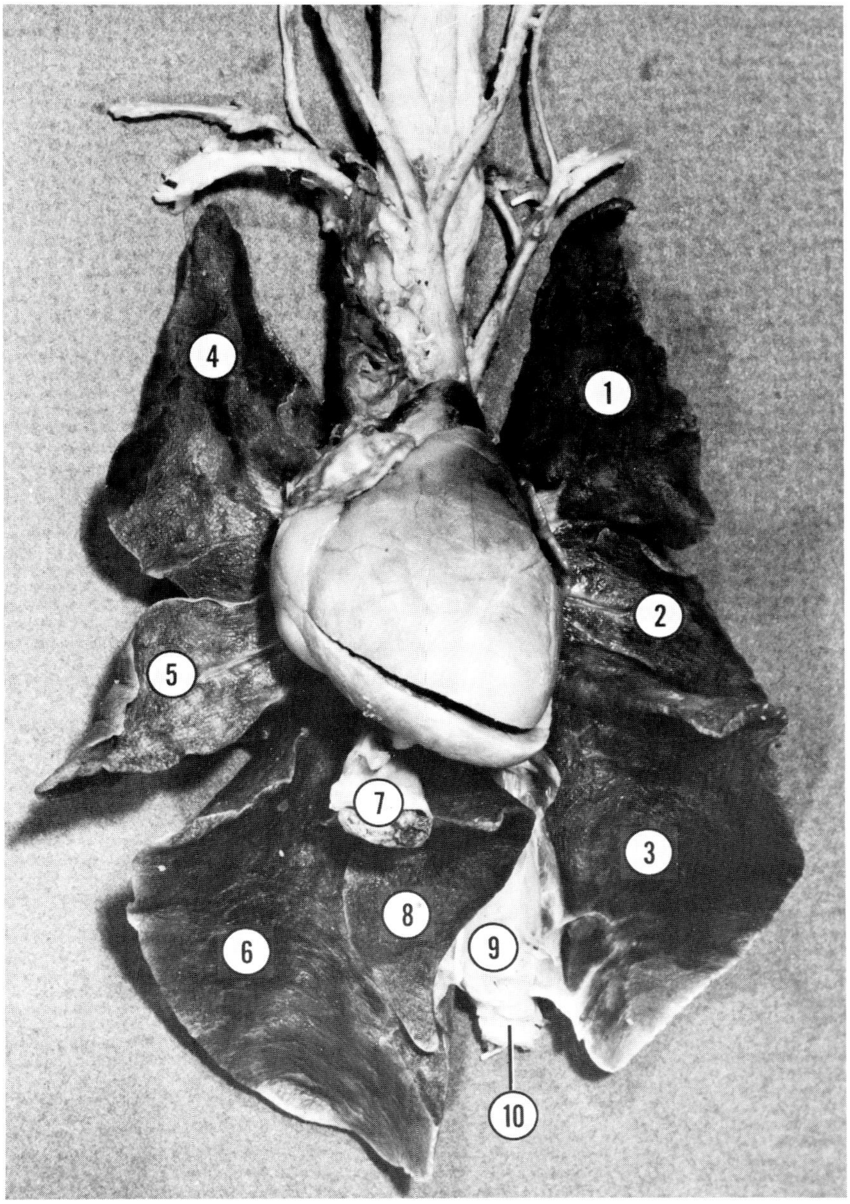

1. **Left anterior lung lobe**
2. **Left middle lung lobe**
3. **Left posterior (Diaphragmatic) lung lobe**
4. **Right anterior lung lobe**
5. **Right middle lung lobe**
6. **Right posterior (Diaphragmatic) lung lobe**
7. **Postcava (Posterior vena cava)**, cut
8. **Intermediate (Mediastinal)** (mē´ dē-AS-ta-nal) **lung lobe**
9. **Esophagus,** cut
10. **Descending thoracic aorta,** cut

FIGURE 49 THORACIC VISCERA REMOVED FROM THORACIC CAVITY TO SHOW RELATIONSHIP BETWEEN RESPIRATORY AND CIRCULATORY STRUCTURES

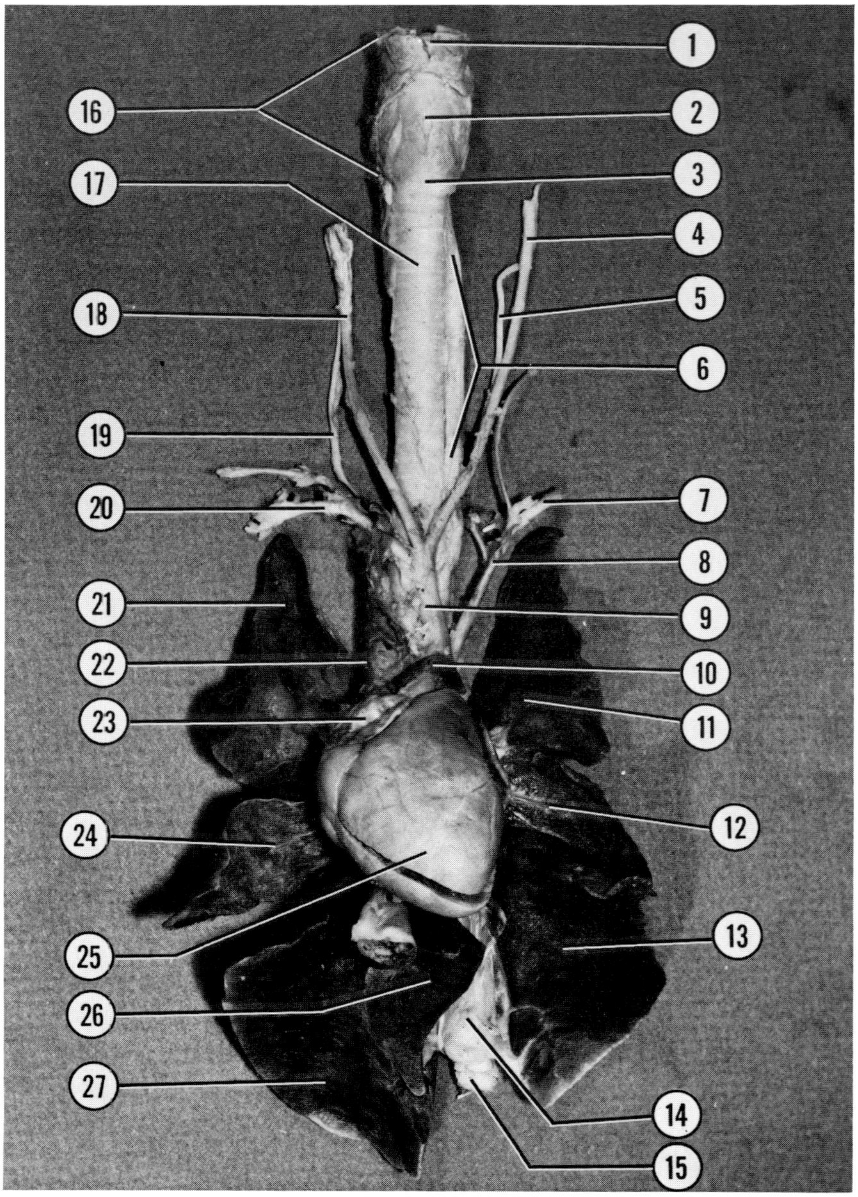

1. **Hyoid** (HĪ-oyd) **bone**
2. **Thyroid cartilage**
3. **Cricoid cartilage**
4. **Left common carotid artery**
5. **Left vagus nerve**
6. **Esophagus**
7. **Left axillary artery**
8. **Left subclavian artery**
9. **Innominate (Brachiocephalic) artery**
10. **Aortic arch**
11. **Left anterior lung lobe**
12. **Left middle lung lobe**
13. **Left posterior (Diaphragmatic) lung lobe**
14. **Esophagus,** cut edge
15. **Descending aorta**
16. **Larynx**
17. **Trachea**
18. **Right common carotid artery**
19. **Right vagus nerve**
20. **Right subclavian artery**
21. **Right anterior lung lobe**
22. **Precava vein (Anterior vena cava)**
23. **Auricle of right atrium**
24. **Right middle lung lobe**
25. **Right ventricle**
26. **Intermediate (Mediastinal) lung lobe**
27. **Right posterior (Diaphragmatic) lung lobe**

FIGURE **50** LUNGS ISOLATED TO SHOW DORSAL VIEW OF TRACHEA, ESOPHAGUS, AND AORTA

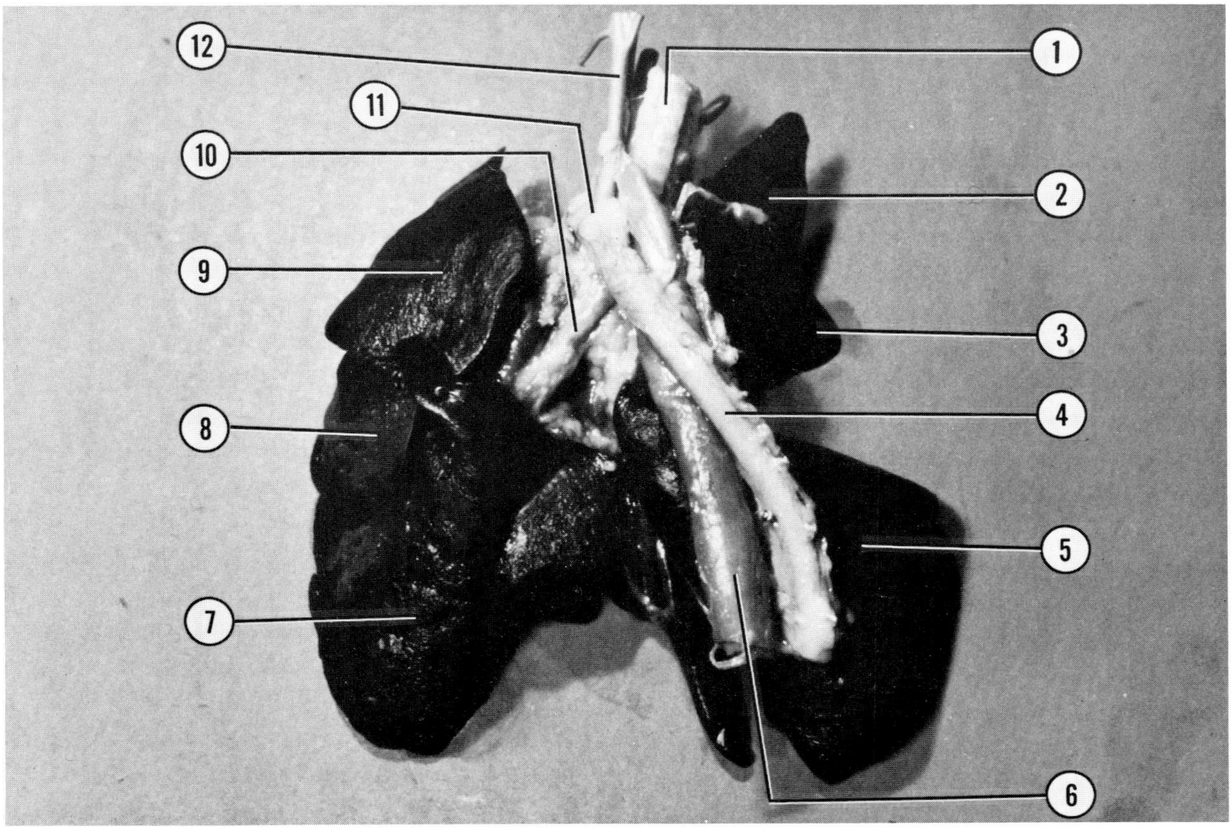

1. **Trachea**
2. **Right anterior lung lobe**
3. **Right middle lung lobe**
4. **Descending thoracic** (thō-RAS-ik) **aorta**
5. **Right posterior lung lobe**
6. **Esophagus**
7. **Left posterior lung lobe**
8. **Left middle lung lobe**
9. **Left anterior lung lobe**
10. **Left primary bronchus** (BRONG-us)
11. **Aortic arch**
12. **Innominate (Brachiocephalic) artery**

FIGURE 51 CORROSION CAST OF ISOLATED LUNGS SHOWING TRACHEA, BRONCHII, AND BRONCHIOLE DETAIL

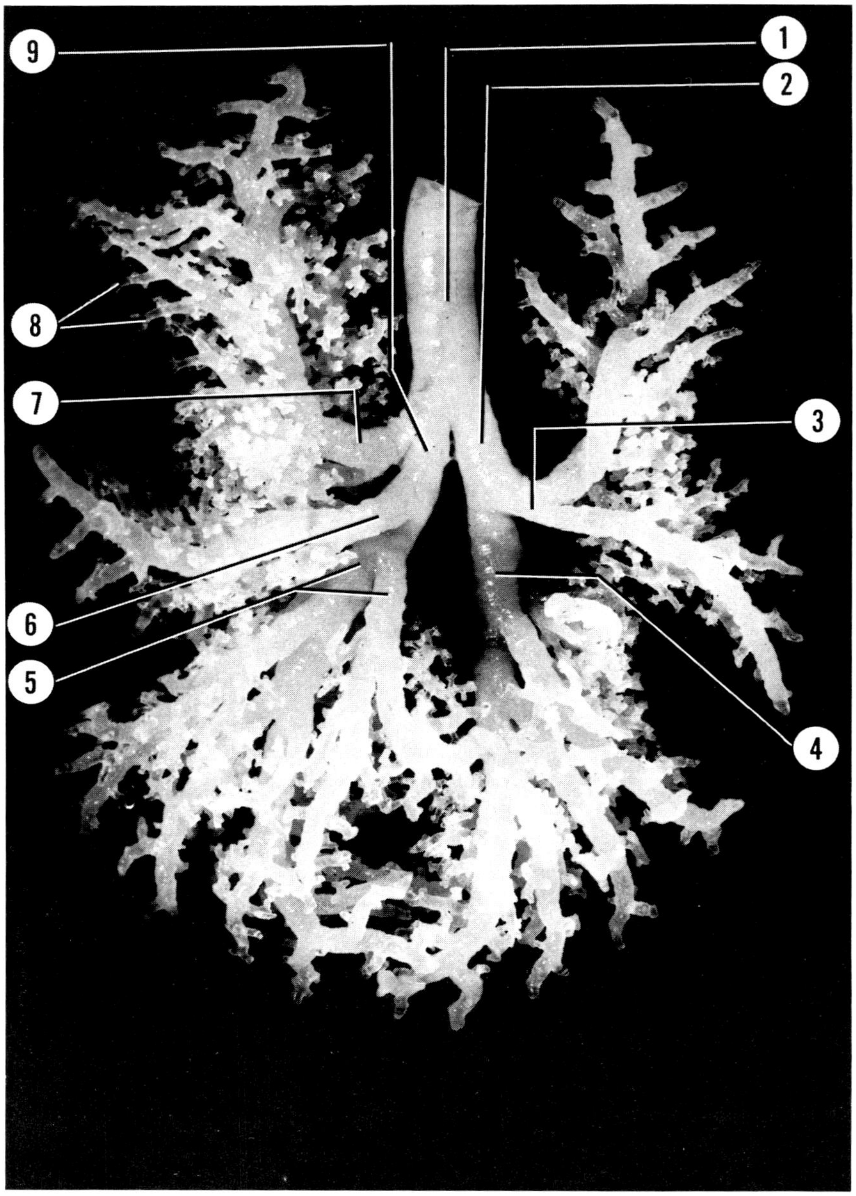

1. **Trachea**
2. **Left primary bronchus**
3. **Left anterior lobar bronchus**
4. **Right posterior lobar bronchus**
5. **Tertiary** (TUR-shē-er ´ē) **bronchus to posterior lung lobe**
6. **Tertiary bronchus to middle lung lobe**
7. **Right anterior lobar bronchus**
8. **Bronchioles** (BRONG-kē-ōls)
9. **Right primary bronchus**

FIGURE **52** ABDOMINAL AND THORACIC VISCERA - VENTRAL VIEW

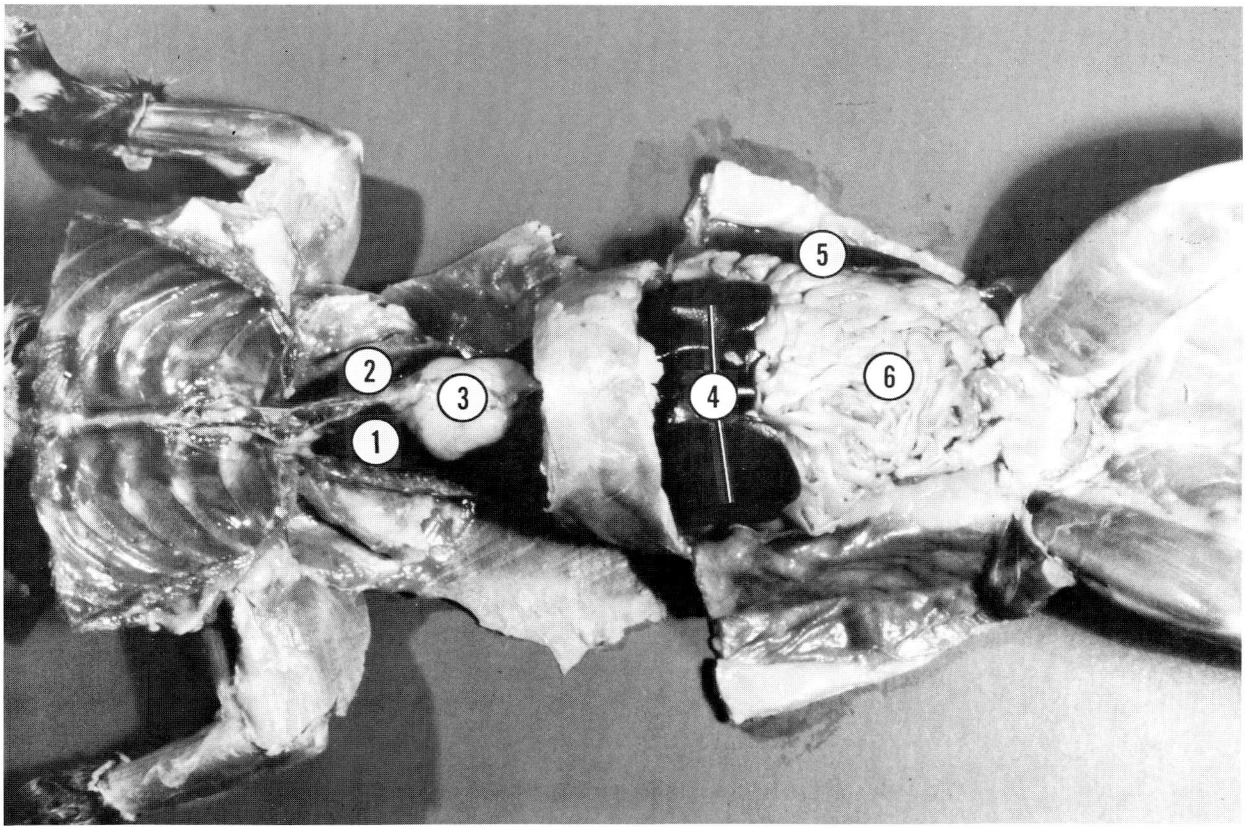

1. **Right lung**
2. **Left lung**
3. **Heart within pericardium** (per´ ē-KAR-dē-um)
4. **Liver**
5. **Spleen**
6. **Greater omentum** (ō-MEN-tum) **containing fat and covering the small intestine**

FIGURE **53** ABDOMINAL VISCERA COVERED BY THE GREATER OMENTUM - VENTRAL VIEW

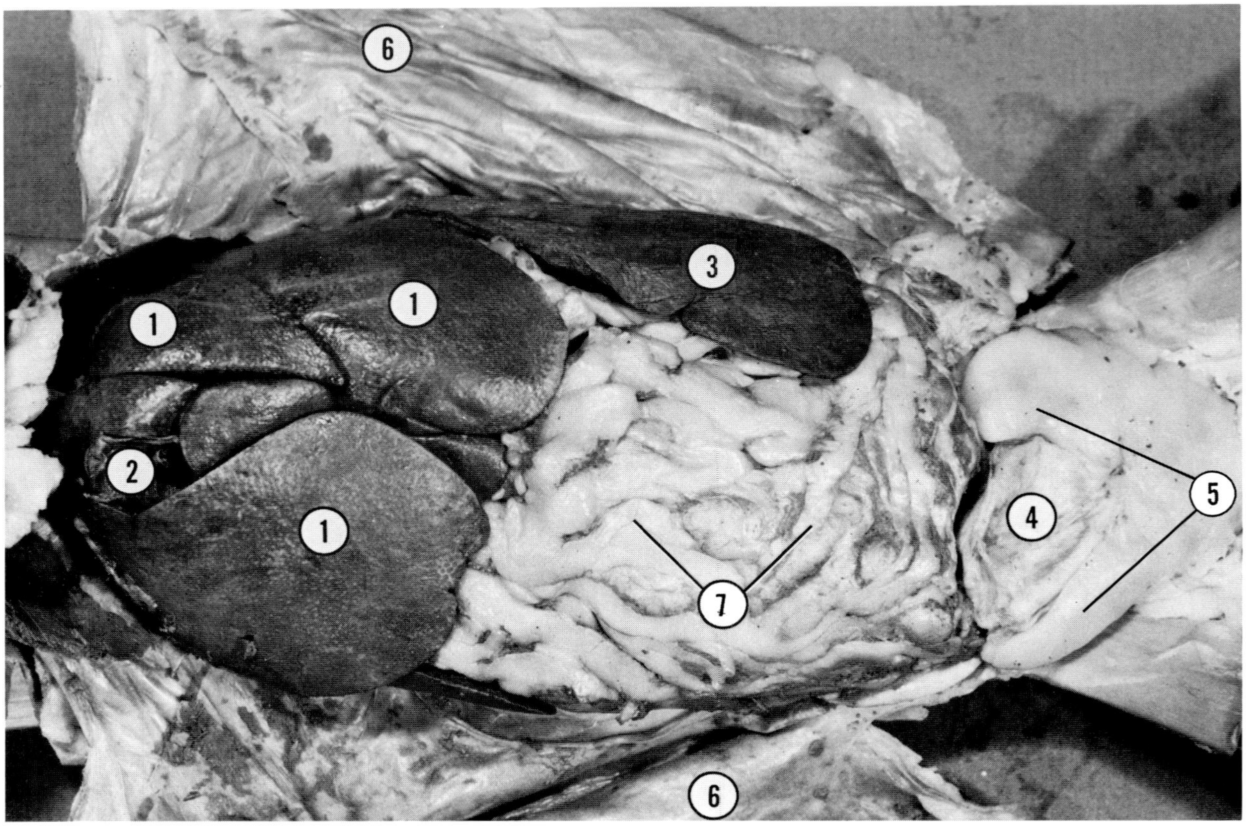

1. **Liver lobes**
2. **Gallbladder**
3. **Spleen**
4. **Urinary** (YOO-ri-ner´ ē) **bladder**
5. **Fat pads**
6. **Abdominal** (ab-DOM-i-nal) **wall showing parietal peritoneum** (pa-RĪ-e-tal per´ i-tō-NĒ-um)
7. **Greater omentum containing fat and covering the small intestine**

FIGURE 54

ABDOMINAL VISCERA VIEWED AFTER REMOVAL OF THE GREATER OMENTUM - VENTRAL VIEW

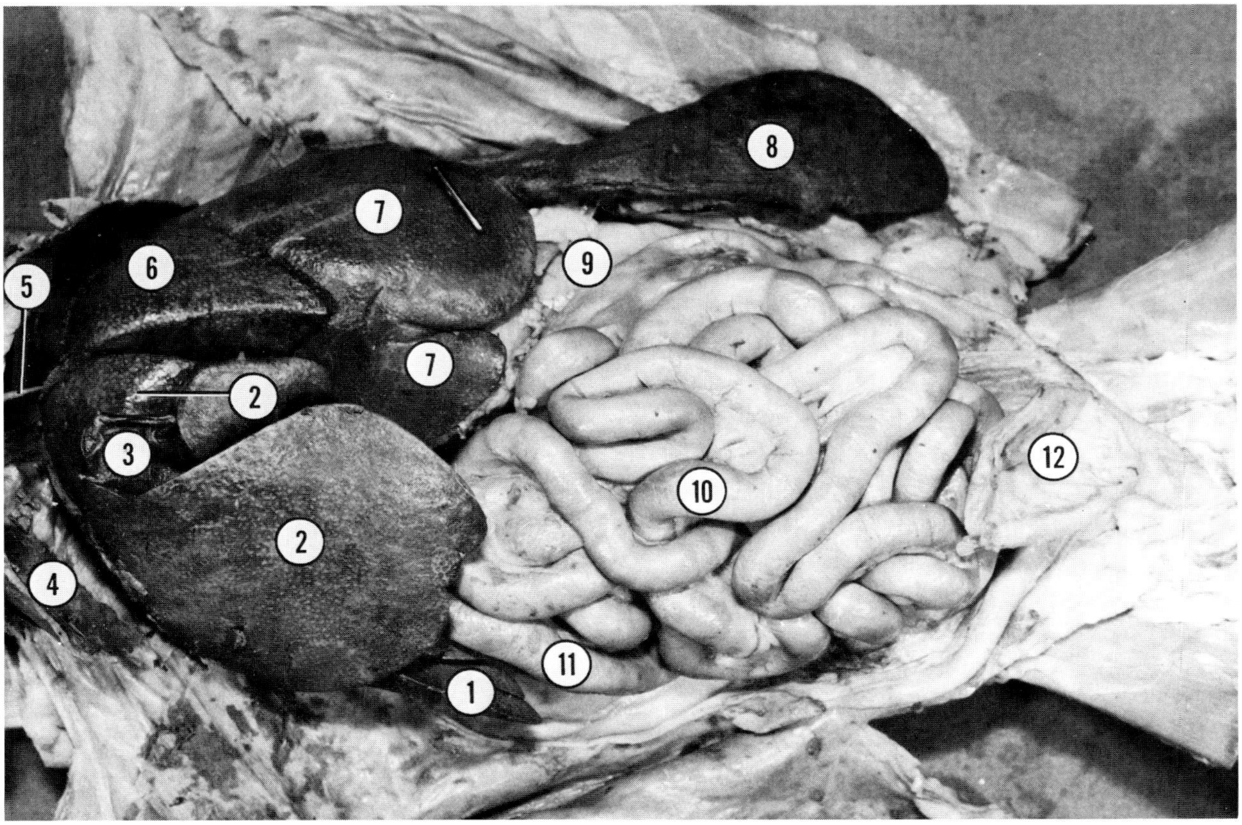

1. **Right lateral liver lobe**
2. **Right medial liver lobe**
3. **Gallbladder**
4. **Diaphragm** (DĪ-a-fram), cut edge
5. **Falciform** (FAL-si-form) **ligament**
6. **Left medial liver lobe**
7. **Left lateral liver lobe**
8. **Spleen**
9. **Stomach,** mostly hidden
10. **Small intestine, jejunum** (je-JOO-num)
11. **Small intestine, duodenum** (doo´ ō-DĒ-num)
12. **Urinary bladder**

FIGURE 55 ABDOMINAL VISCERA - RIGHT LATERAL VIEW (GREATER OMENTUM REMOVED)

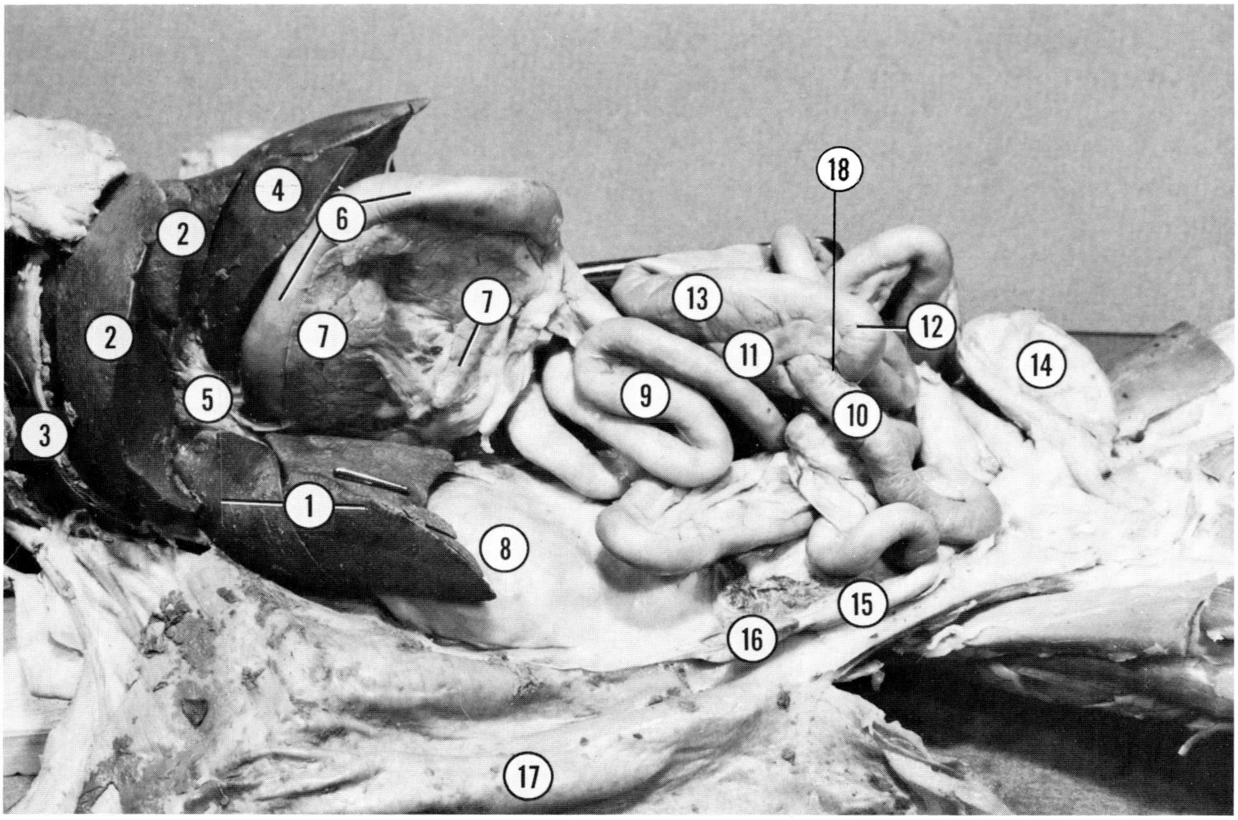

1. **Right lateral liver lobe**
2. **Right medial liver lobe**
3. **Diaphragm,** cut edge
4. **Left lateral liver lobe**
5. **Common bile duct and hepatic** (he-PAT-ik) **portal vein**
6. **Duodenum**
7. **Pancreas** (PAN-krē-us)
8. **Kidney covered with peritoneum**
9. **Jejunum**
10. **Ileum** (IL-ē-um)
11. **Lymph nodes**
12. **Cecum** (SĒ-kum)
13. **Ascending colon** (KŌ-lon)
14. **Urinary bladder**
15. **Right uterine** (YOO-ter-in) **horn**
16. **Right ovary** (Ō-var-ē)
17. **Abdominal wall**
18. **Ileocecal** (il-ē-ō-SĒ-kal) **valve**

FIGURE 56 INTERNAL DETAIL OF THE STOMACH

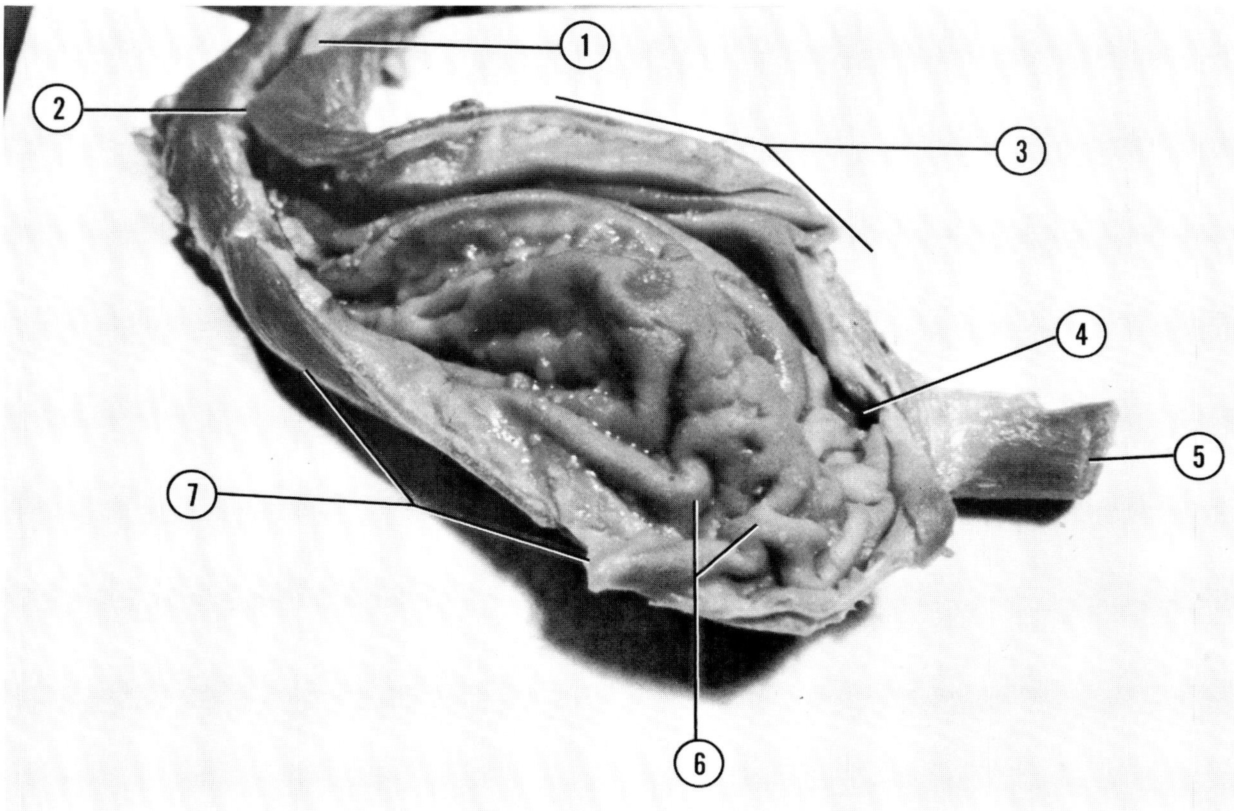

1. **Duodenum**
2. **Pyloric** (pī-LOR-ik) **valve region**
3. **Body region of stomach, reflected wall of the greater curvature**
4. **Cardiac** (KAR-dē-ak) **orifice**
5. **Esophagus** (ē-SOF-a-gus)
6. **Rugae** (ROO-gā) **in fundus region**
7. **Greater curvature, body region of stomach**

FIGURE **57** ABDOMINAL VISCERA SHOWING DETAIL OF LIVER AND DUODENUM - RIGHT VENTRAL VIEW

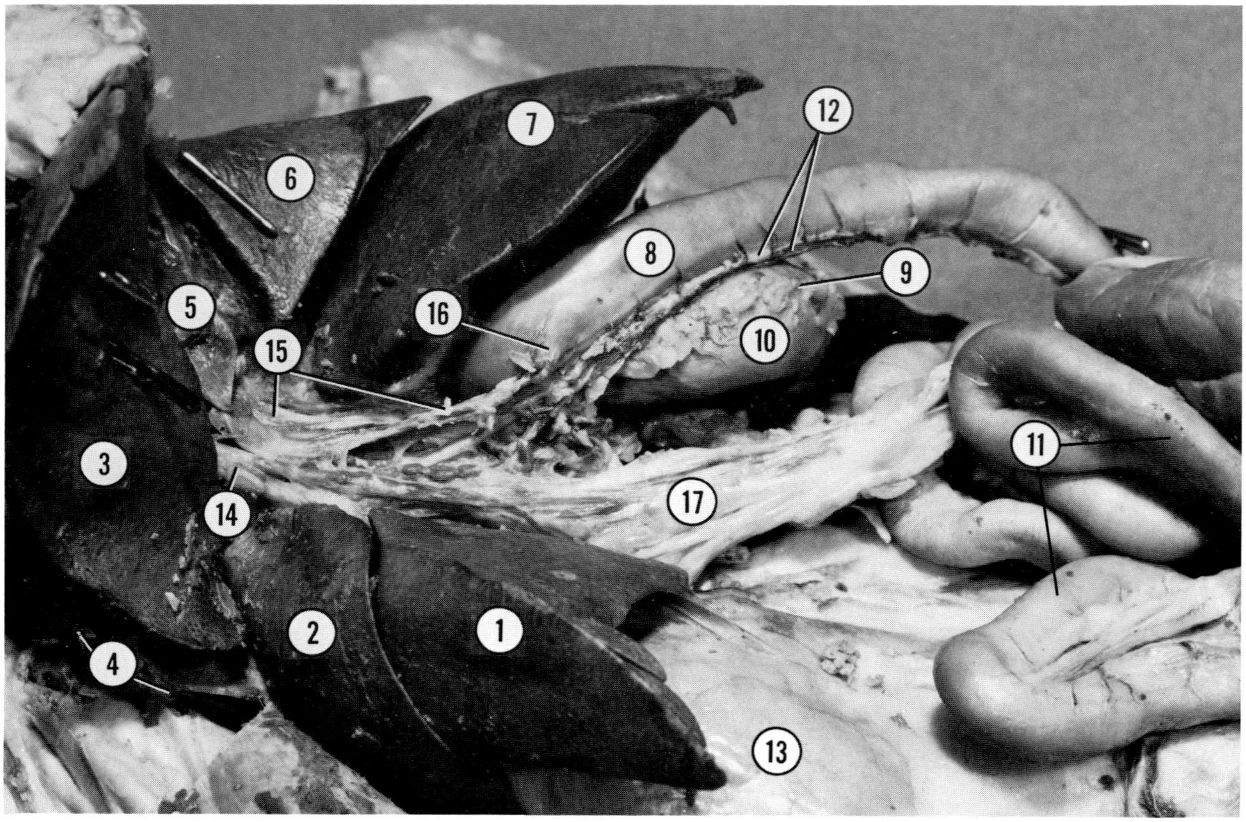

1. **Right lateral posterior liver lobe**
2. **Right lateral liver lobe**
3. **Right medial liver lobe**
4. **Diaphragm**
5. **Gallbladder**
6. **Left medial liver lobe**
7. **Left lateral liver lobe**
8. **Duodenum**
9. **Greater omentum,** cut edge
10. **Greater curvature of stomach**
11. **Jejunum**
12. **Anterior pancreaticoduodenal** (pan´ krē-AT-ik-Ō-doo´ ō-DĒ-nal) **artery**
13. **Right kidney with fat capsule**
14. **Hepatic portal vein**
15. **Common bile duct**
16. **Ampulla of Vater** (am-PYOO-la va-TEER)
17. **Mesoduodenum** (mez´ ō-doo´ ō-DĒ-num)

FIGURE **58** THE STOMACH AND PANCREAS

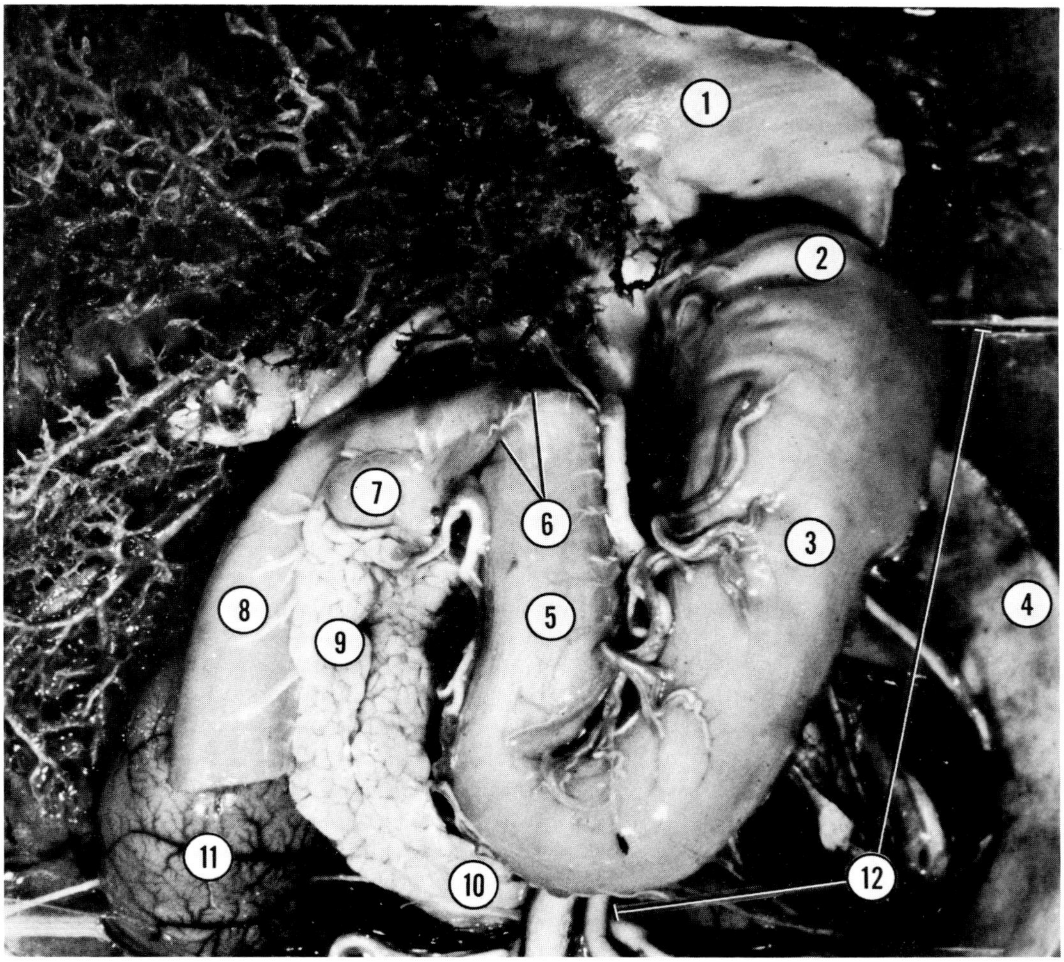

1. **Diaphragm**
2. **Fundic** (FUN-dik) **portion of stomach**
3. **Body of stomach**
4. **Spleen**
5. **Pyloric region of stomach**
6. **Pyloric valve**
7. **Lymph node**
8. **Duodenum**
9. **Duodenal** (doo´ ō-DĒ-nal) **portion of pancreas**
10. **Gastrosplenic** (gas´ trō-SPLEN-ik) **portion of pancreas**
11. **Right kidney**
12. **Greater curvature of stomach**

FIGURE 59 THE UROGENITAL ORGANS - VENTRAL VIEW

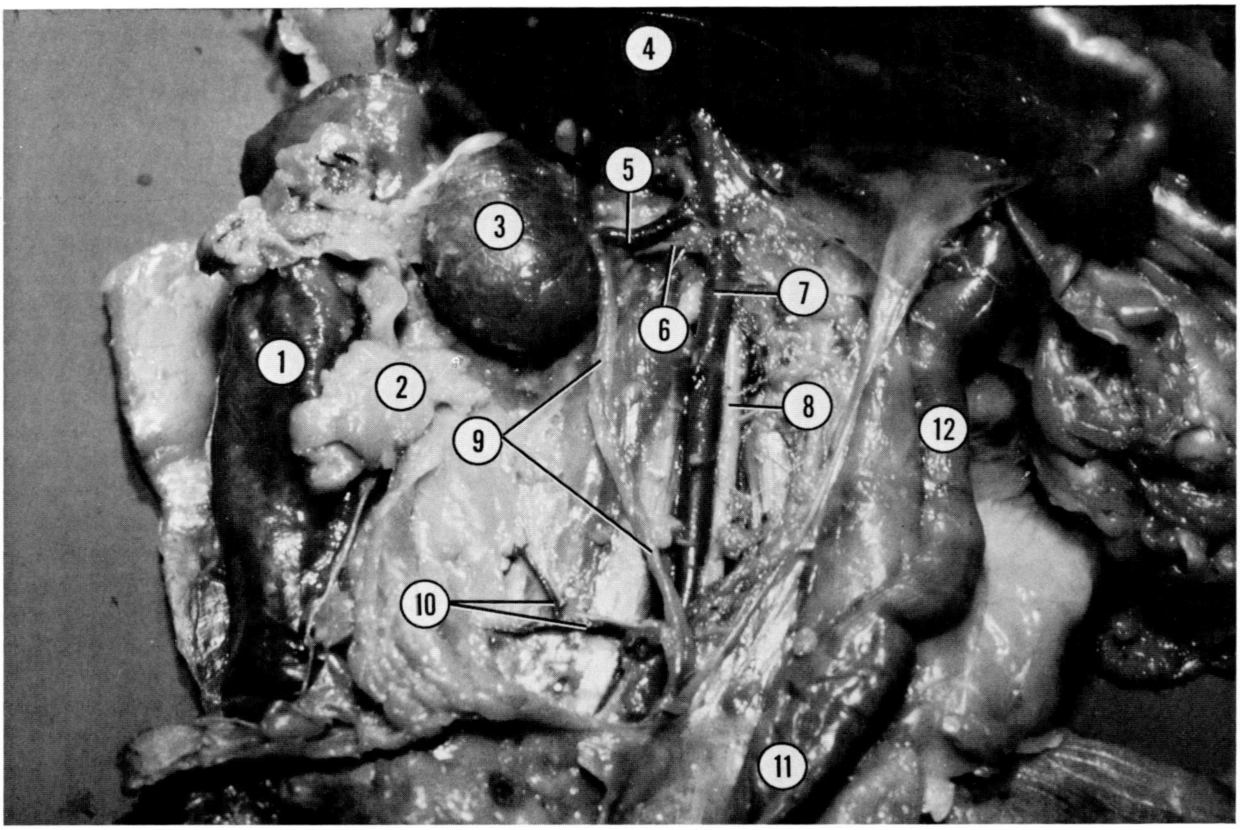

1. **Body wall**
2. **Perirenal** (per-ē-RĒ-nal) **fat capsule removed from kidney**
3. **Right kidney**
4. **Liver**
5. **Right renal** (RĒ-nal) **vein**
6. **Right renal artery**
7. **Posterior vena cava** (VĒ-na KĀ-va)
8. **Abdominal aorta** (ab-DOM-i-nal ā-OR-ta)
9. **Right ureter** (YOO-re-ter)
10. **Right deep circumflex iliac** (SUR-kum-flex IL-ē-ak) **artery and vein**
11. **Rectum** (REK-tum)
12. **Descending colon**

FIGURE 60 FRONTAL SECTION OF THE SHEEP KIDNEY

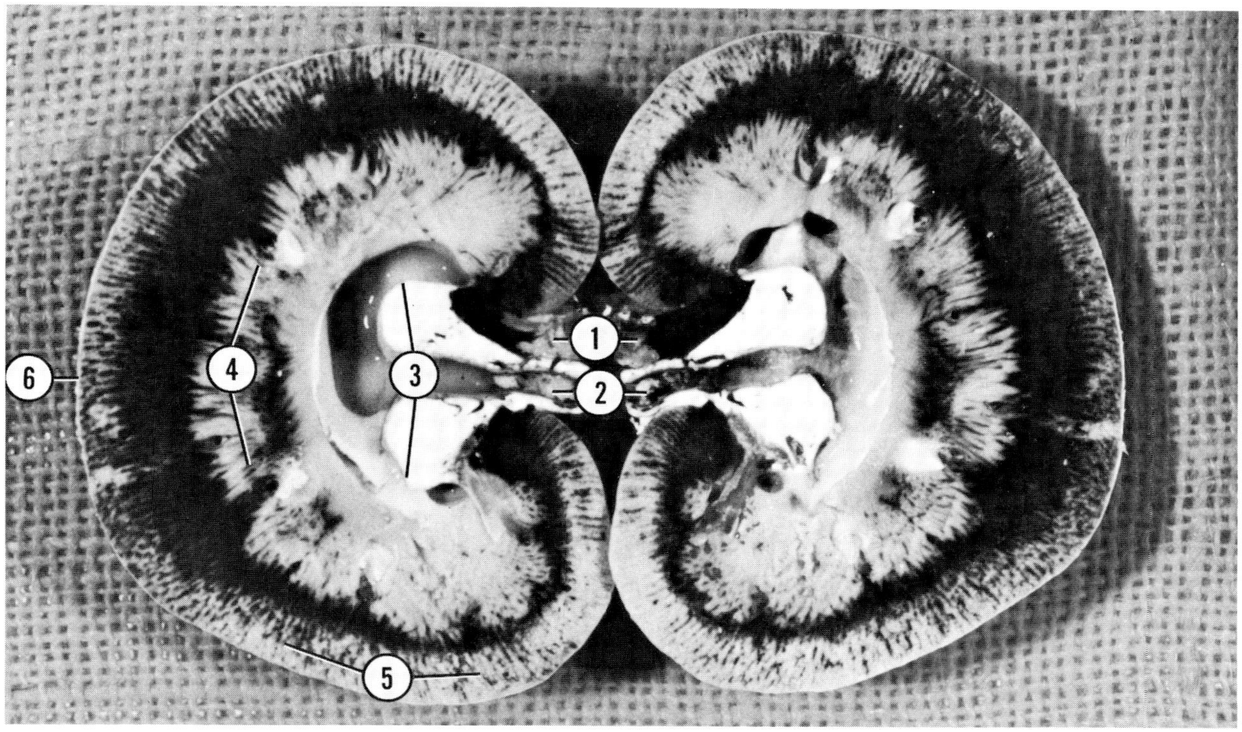

1. **Renal** (RĒ-nal) **artery**
2. **Ureter** (YOO-re-ter)
3. **Renal pelvis**
4. **Medulla** (me-DULL-la)
5. **Cortex** (KOR-teks) **region**
6. **Renal capsule**

FIGURE 61 FRONTAL SECTION OF THE SHEEP KIDNEY SHOWING INTERNAL STRUCTURAL DETAIL

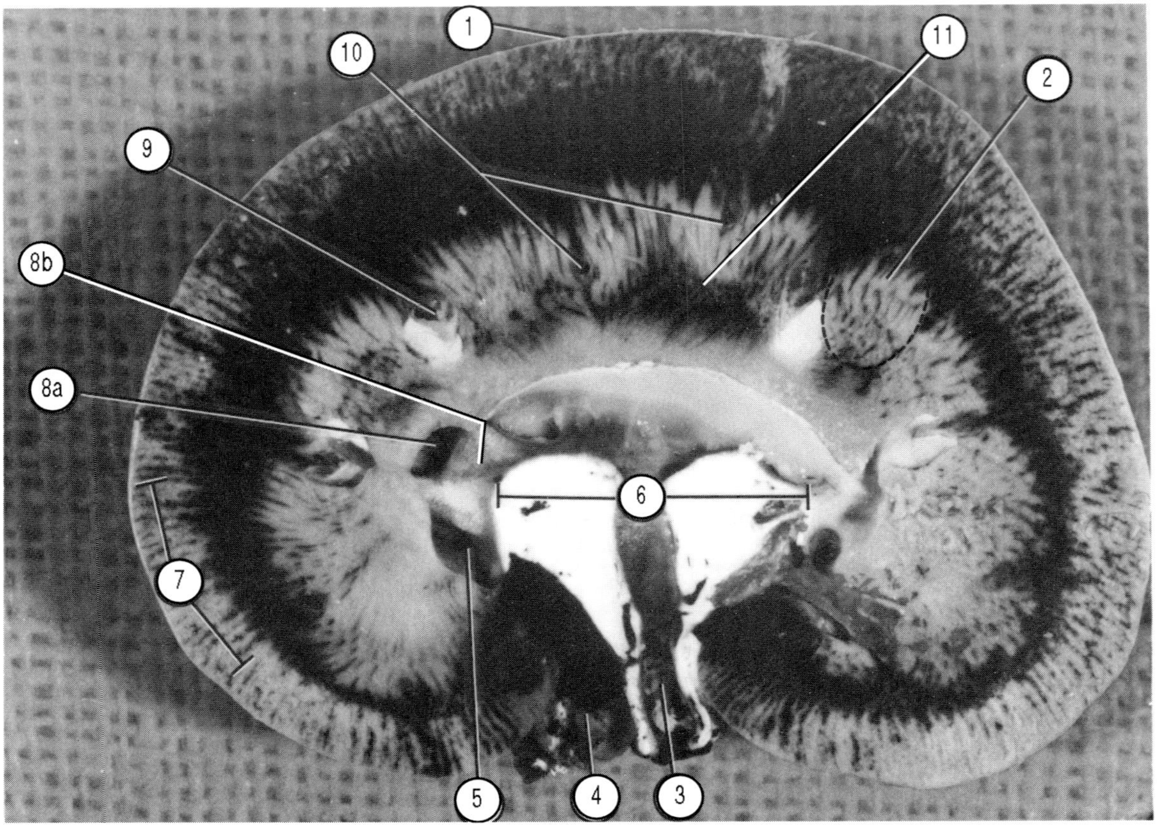

1. **Renal capsule**
2. **Renal pyramid**
3. **Ureter**
4. **Renal artery**
5. **Renal sinus**
6. **Renal pelvis**
7. **Cortex region**

8a. **Minor calyx** (KĀ-lix)
8b. **Major calyx**

9. **Renal column**
10. **Arcuate** (AR-kū-āt) **artery and vein**
11. **Renal papilla** (pa-PIL-ah)

FIGURE **62** SHEEP KIDNEY SHOWING HILUS REGION DETAIL-MEDIAL VIEW

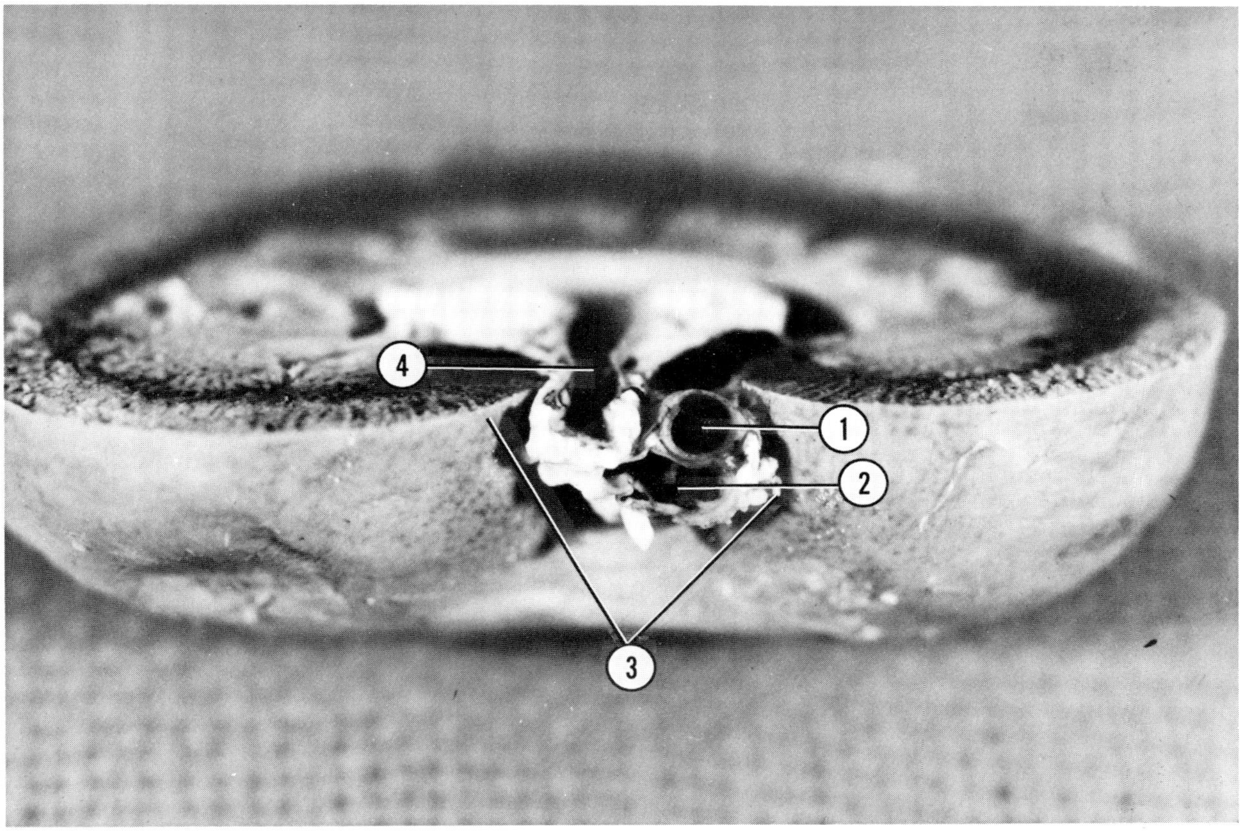

1. **Renal artery**
2. **Renal vein**
3. **Hilus** (HĪ-lus) **region**
4. **Ureter**

FIGURE 63 CORROSION PREPARATION OF A SHEEP KIDNEY SHOWING DETAIL OF VASCULARIZATION

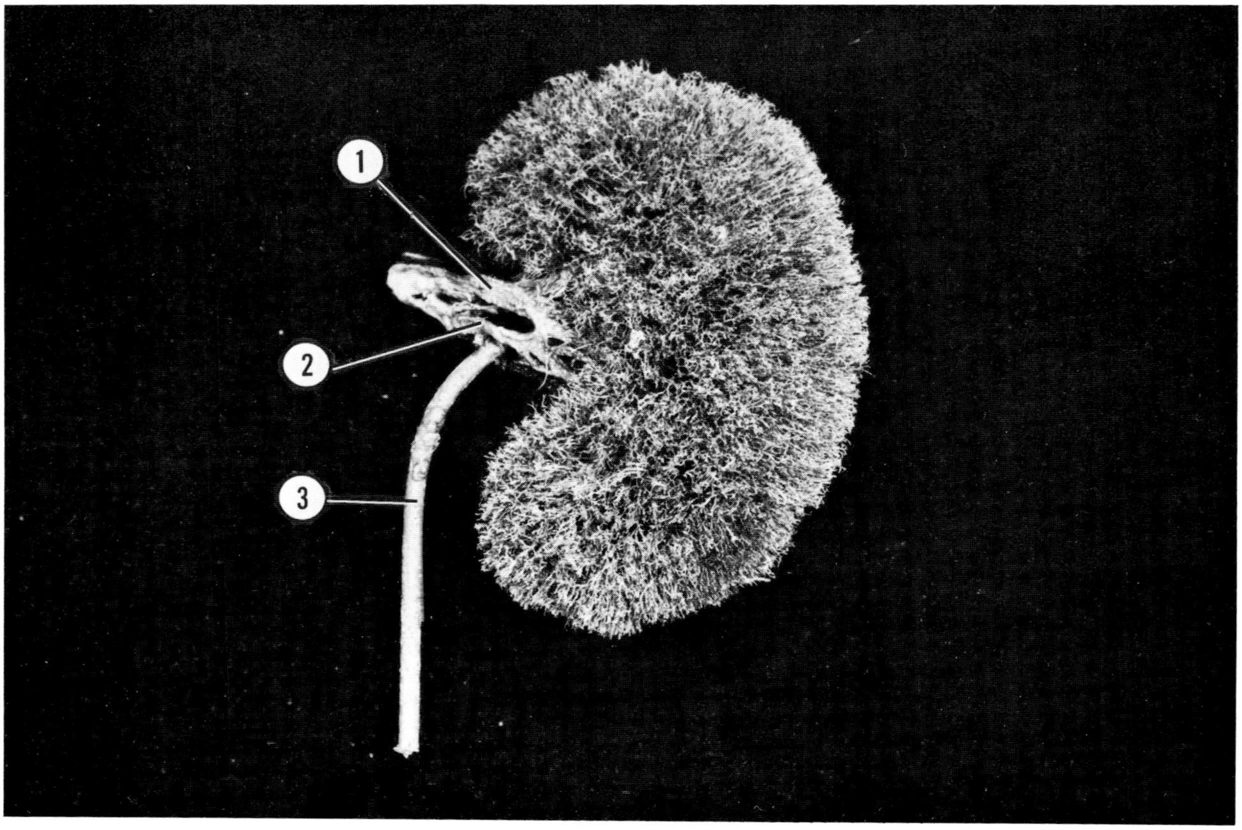

1. **Renal vein**
2. **Renal artery**
3. **Ureter**

FIGURE 64 PANORAMIC VIEW OF THE ABDOMINAL VISCERA AND FEMALE UROGENITAL ORGANS

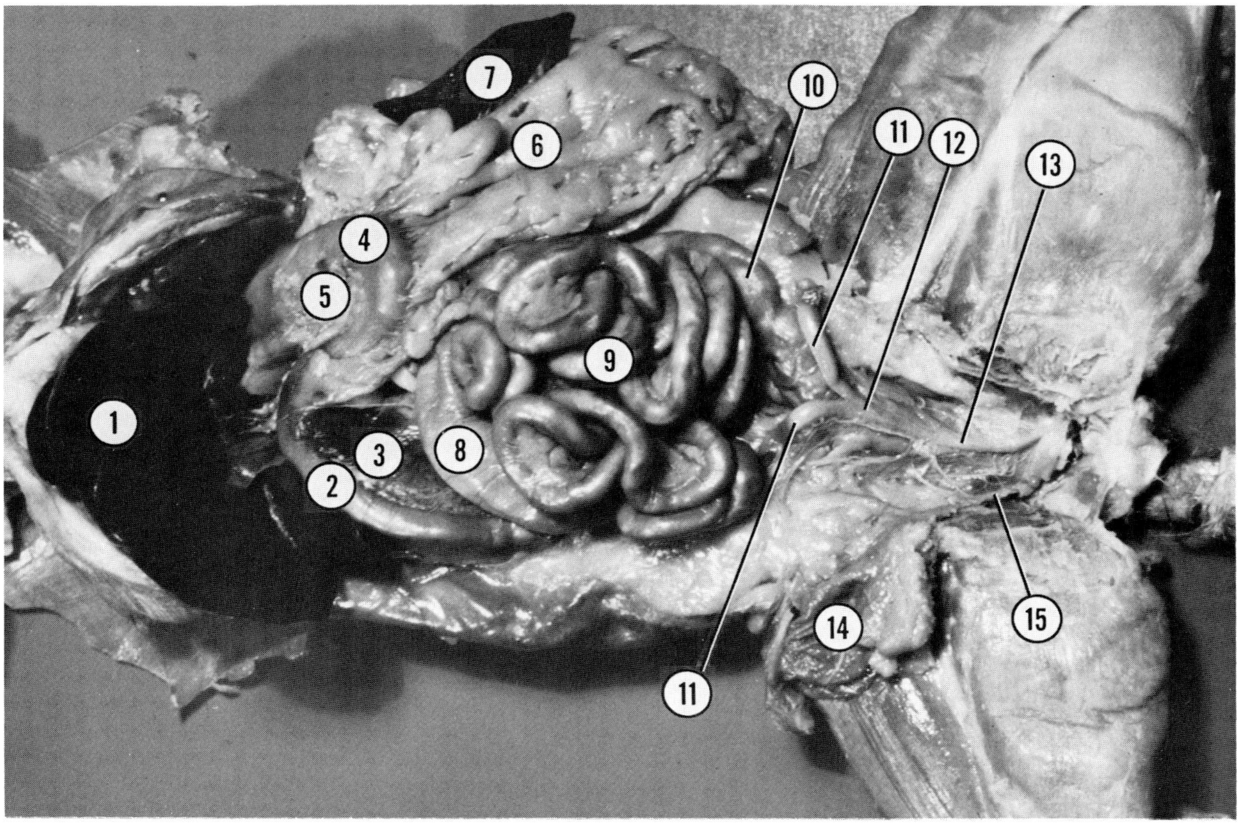

1. **Liver**
2. **Duodenum**
3. **Pancreas**
4. **Stomach**
5. **Lesser omentum**
6. **Greater omentum**
7. **Spleen**
8. **Ascending colon**
9. **Small intestine**
10. **Descending colon**
11. **Uterine** (YOO-ter-in) **horns**
12. **Body of the uterus** (YOO-te-rus)
13. **Vagina** (va-JĪ-na)
14. **Urinary** (YOO-ri-ner´ ē) **bladder**
15. **Urethra** (yoo-RĒ-thra)

FIGURE 65

FEMALE UROGENITAL ORGANS WITHIN THE PELVIC CAVITY - VENTRAL VIEW (VENTRAL PORTION OF OSSA COXAE CUT AND REFLECTED)

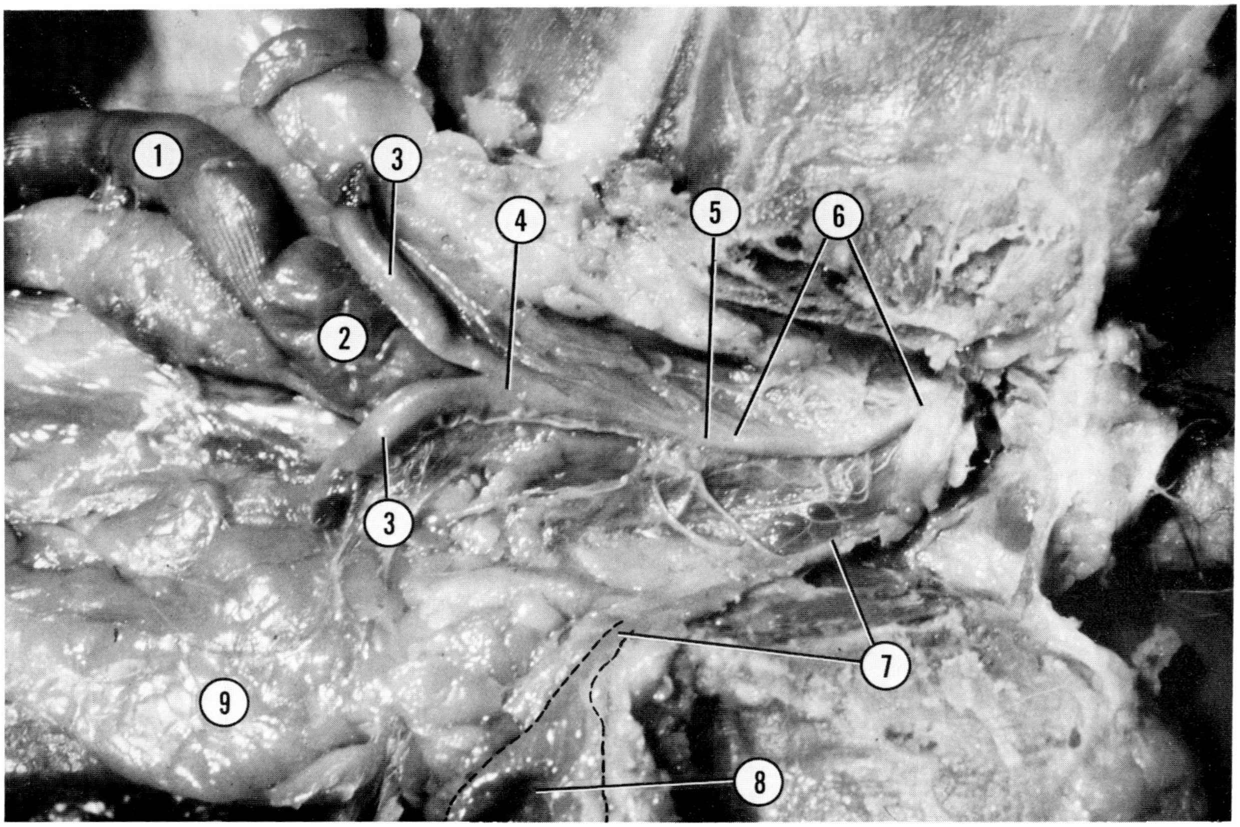

1. **Descending colon**
2. **Rectum**
3. **Uterine horn**
4. **Body of uterus**
5. **Cervix** (SER-viks) **region**
6. **Vagina**
7. **Urethra**
8. **Urinary bladder**
9. **Body fat**

FIGURE 66

VENTRAL VIEW OF FEMALE UROGENITAL ORGANS ISOLATED FROM ABDOMINAL CAVITY

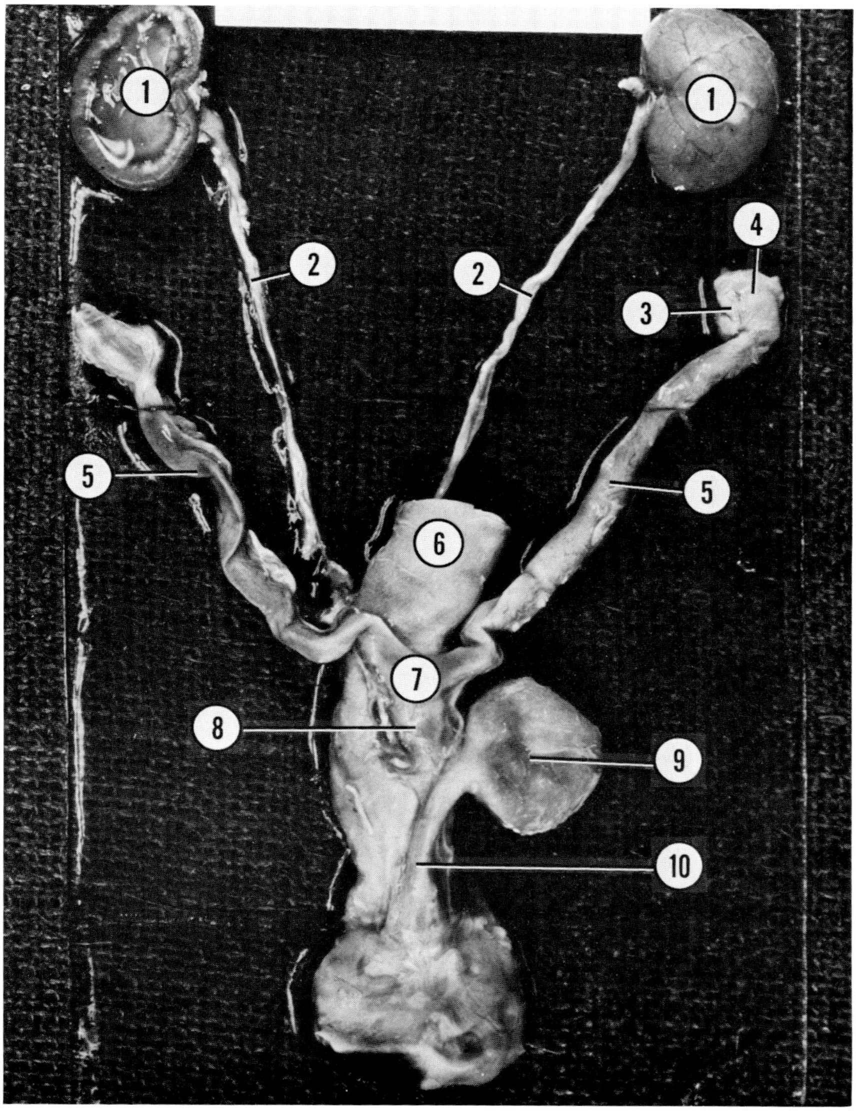

1. **Kidneys**
2. **Ureters**
3. **Left fallopian** (fa-LŌ-pē-un) **tube**
4. **Left ovary** (Ō-var-ē)
5. **Uterine horn**
6. **Rectum**
7. **Body of uterus**
8. **Vagina**
9. **Urinary bladder**
10. **Urethra**

FIGURE 67

VENTRAL VIEW OF ABDOMINAL CAVITY OF A FULL-TERM CAT EXPOSED TO SHOW UTERINE HORNS CONTAINING FETUSES

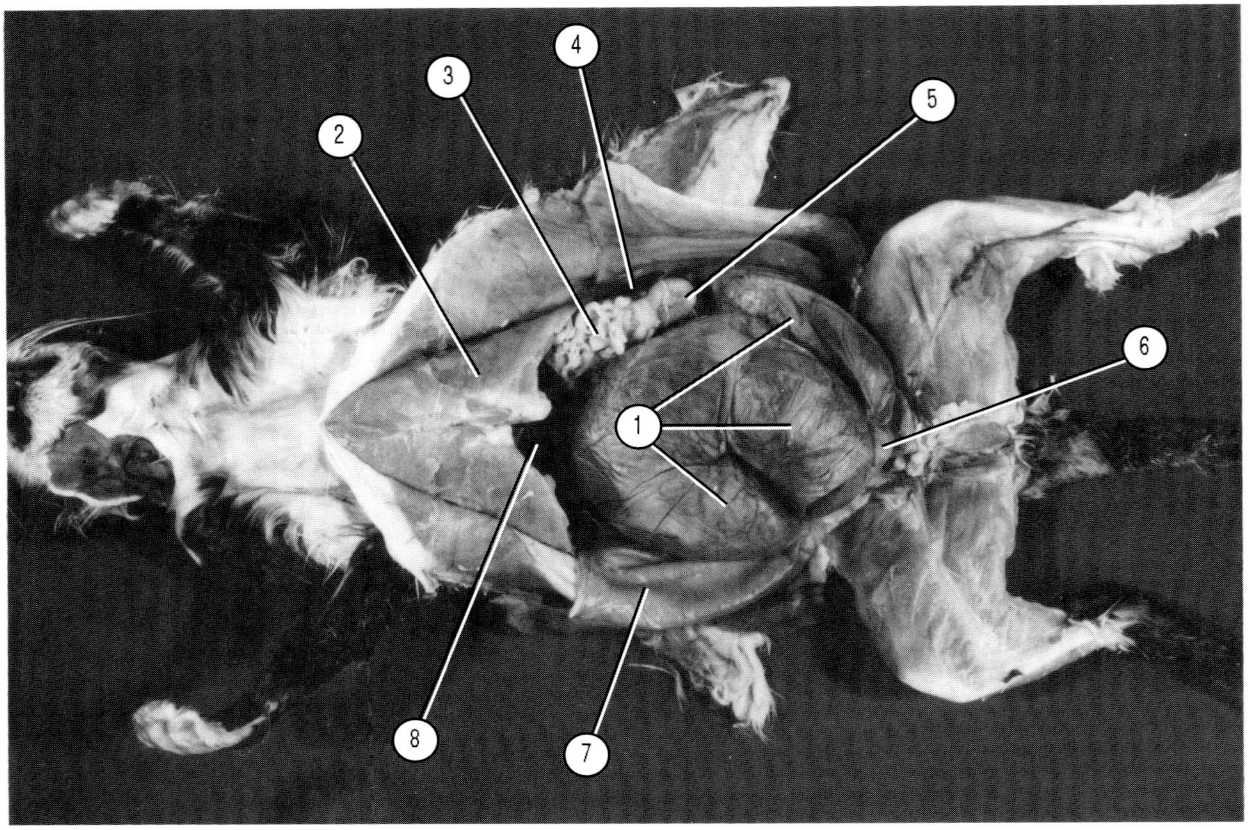

1. **Uterine horns containing fetuses** (FĒ-ta-suz)
2. **Thorax**
3. **Greater omentum over stomach and intestine**
4. **Spleen**
5. **Jejunum**
6. **Urinary bladder**
7. **Body wall**
8. **Liver, right medial lobe**

FIGURE 68 ISOLATED MALE UROGENITAL ORGANS - VENTRAL VIEW

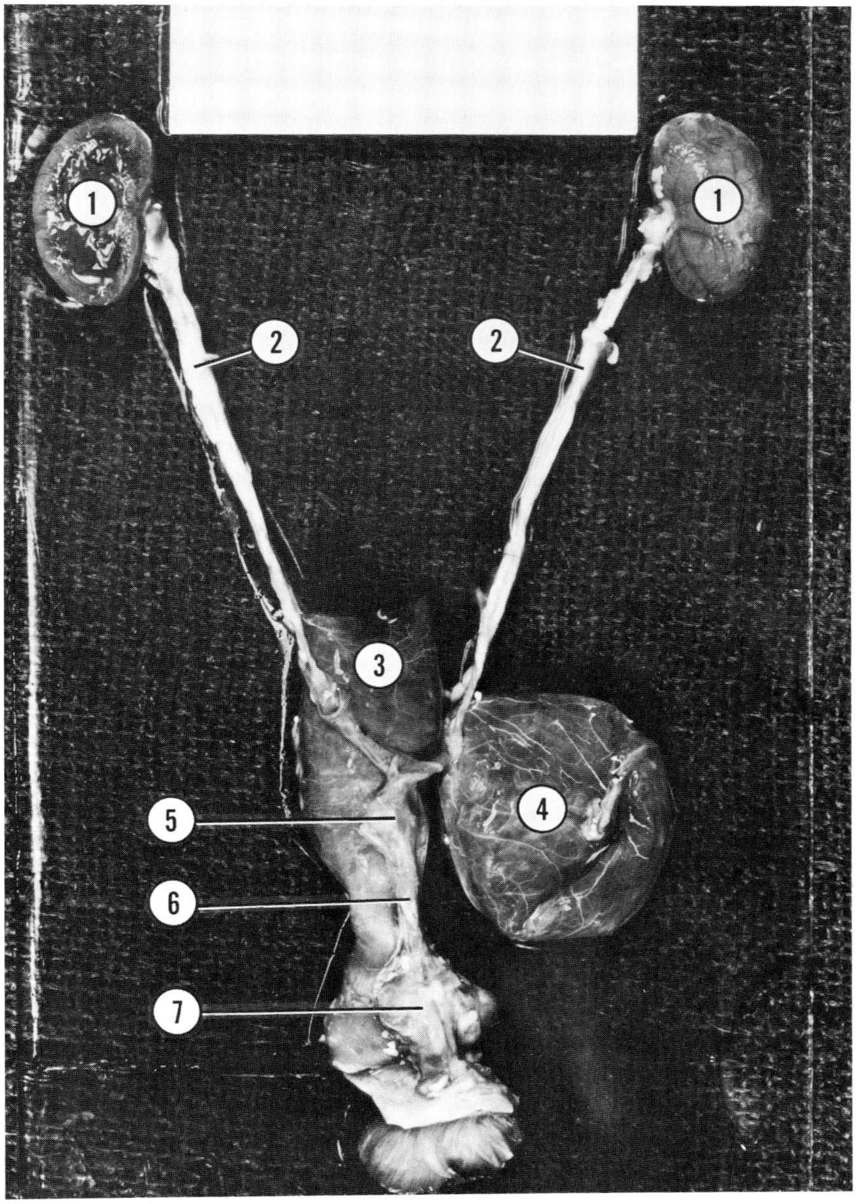

1. **Kidneys**
2. **Ureters** (YOO-re-ters)
3. **Rectum**
4. **Urinary** (YOO-ri-ner´ ē) **bladder**
5. **Prostate** (PROS-tāt) **gland**
6. **Male urethra** (yoo-RĒ-thra)
7. **Bulbourethral** (bul´ bō-yoo-RĒ-thral) **gland**

FIGURE **69**

ISOLATED MALE UROGENITAL SYSTEM SHOWING EXTERNAL REPRODUCTIVE ORGAN DETAIL - VENTRAL VIEW

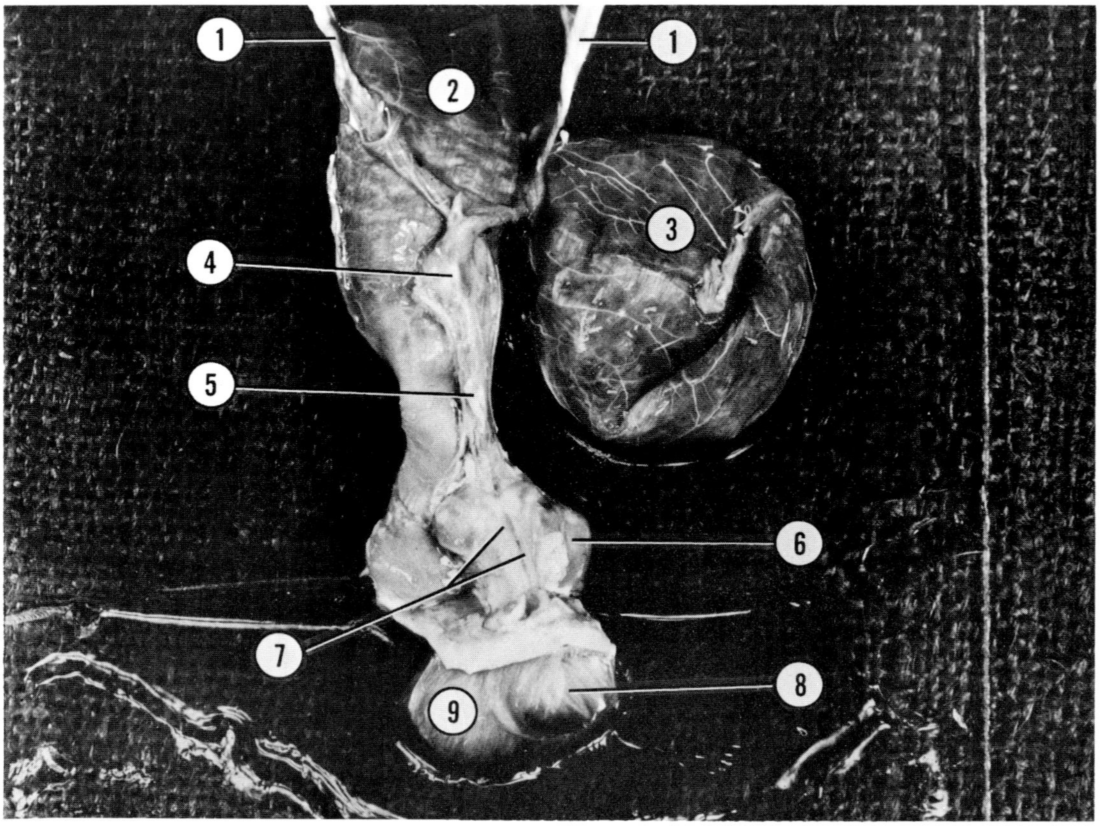

1. **Ureters**
2. **Rectum**
3. **Urinary bladder**
4. **Prostate gland**
5. **Male urethra**
6. **Bulbourethral** (bul´ bō-yoo-RĒ-thral) **gland**
7. **Crus of penis** (krus PĒ-nis)
8. **Penis**
9. **Testis** (TES-tis) **within the scrotum** (SKRŌ-tum)

Index

For convenience the numbers that follow each term are for "figure numbers," not page numbers.

LYMPHATIC SYSTEM

ARTERIES

VEINS

RESPIRATORY SYSTEM

DIGESTIVE SYSTEM

UROGENITAL SYSTEM

KIDNEY